THE STICKLER'S GUIDE

TO SCIENCE IN THE AGE

OF MISINFORMATION

Published in 2021 by Timber Press, Inc.

The Haseltine Building

133 S.W. Second Avenue, Suite 450

Portland, Oregon 97204-3527

timberpress.com

Printed in the United States of America

Text design by Vincent James

Cover design by Mark Melnick

Interior illustrations by Johnny Bertram

ISBN 978-1-64326-042-6

Catalog records for this book are available
from the Library of Congress and the British Library.

THE STICKLER'S GUIDE TO SCIENCE IN THE AGE OF MISINFORMATION

The Real Science Behind Hacky Headlines, Crappy Clickbait, and Suspect Sources

R. PHILIP BOUCHARD

Timber Press | Portland, Oregon

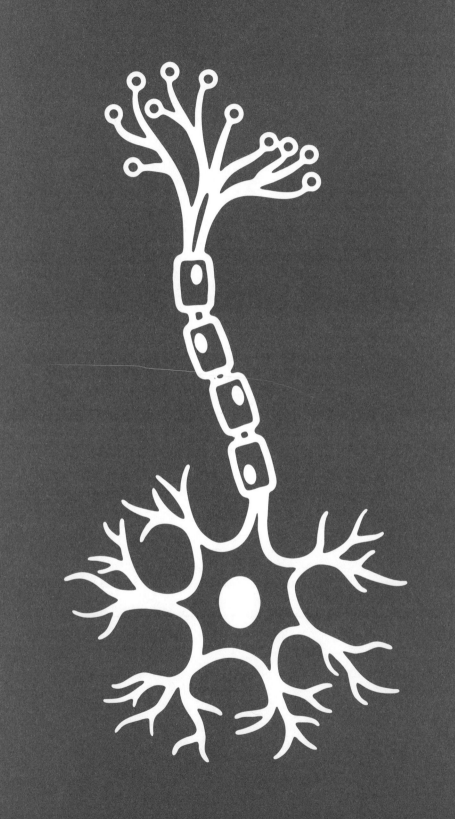

CONTENTS

PREFACE

Here's a quick quiz.

Which of the following statements are true?

- ☐ DNA is the blueprint of life.
- ☐ Rainforests are the lungs of the planet.
- ☐ There is no gravity in space.

If you see this as a trick question, you're right. Although the third statement (which reflects a popular misconception about gravity) is objectively false, the first two statements are more difficult to evaluate. Both rely on metaphors instead of presenting objective facts, so they're not easily categorized as either true or false. Yet both statements are frequently presented as valuable truths, encapsulating important knowledge about science. To evaluate either metaphor, we first have to interpret it—to restate it in terms of objective assertions. This is exceedingly tricky, because we can easily draw wildly different conclusions from either metaphor.

The problem here is not just that metaphors are indirect. The broader issue is our tendency to rely on short, familiar phrases to represent large chunks of knowledge. Consider the following examples:

- ☐ the five senses
- ☐ survival of the fittest
- ☐ killing germs
- ☐ high levels of radiation
- ☐ twenty-four hours a day
- ☐ full of energy
- ☐ a left-brained person
- ☐ global warming

Each of these phrases is intended to be easily understood, encapsulating a great deal of information we think we already know. In essence, each of these phrases is a kind of meme. In our mass media and social networks, we depend on these verbal shortcuts to make our communication more efficient. Because the phrases are so familiar, we tend to feel confident in our understanding of what these phrases mean. Yet each of these phrases provides cover for common scientific misconceptions. More often than not, our understanding of the science behind these phrases is an awkward blend of truth and untruth. It's not that our popular beliefs about science are all wrong; it's that they are often not quite right.

That leads to the first goal of this book: to look at the real science behind popular phrases such as these. In other words, what is it we think we know that isn't quite what we think?

The second goal is to connect the dots. We live in the Information Age, constantly bombarded by disconnected bits of information that are promoted with great urgency by countless media sources and social contacts. When all that information remains disconnected, it's merely a collection of trivia without a larger meaning. It's like standing 2 inches from a pointillist painting and staring at the individual dots. I admit that looking at the individual dots—those bits of trivia—can sometimes be fascinating. But the real value comes when you can finally see the big picture, which is not until you step back to see the entire canvas at once. To me, all of those little factoids lack significance until I can see how they fit together.

In my experience, most people are truly interested in connecting the dots—assembling a coherent picture from selected bits of information. But unfortunately, not all of the factoids we encounter are truly factual. The result can be like a huge jumble of puzzle pieces, half of which are decoys that don't actually fit into the big picture. If you have trouble choosing and assembling the correct puzzle pieces, the big picture might never come into view. Or worse yet, a completely erroneous big picture might emerge. In the thirteen chapters of this book, I sort through a prodigious pile of puzzle pieces and select the useful ones that can actually contribute to an accurate, connected picture.

My third and final goal with this book is simply to have a bit of fun. Science information can be presented in a way that's dull and boring, or it can be presented in a way that's lively and interesting. I certainly want to present my collection of dots and puzzle pieces in a manner that you will find highly engaging—to give you a book that's truly enjoyable to read. In short, this book is intended to be a fun exercise in connecting the dots that lurk behind the common shorthand phrases we use when we talk about science. Wait—I don't think I've quite captured it. Perhaps I should just say that you and I are about to work together on a big jigsaw puzzle, and I hope to be a lively raconteur as we play.

But before we get started, I must offer a warning. Although I am a stickler for science details, eager to get each of my facts exactly right, you should not assume I've actually gotten everything perfectly correct. Part of the issue is that science is actually a process, not a collection of facts. This process constantly generates new information, resulting in the continuous questioning of old assumptions. Our knowledge of science undergoes a never-ending process of refinement and reinterpretation of the details. This constant change allows us to see the big picture with ever more clarity, and once in a while it even causes a noteworthy shift in the appearance of the big picture. So there's no way I'm ever going to be 100 percent correct. Plus, I've tried to cover a wide range of topics, and my understanding of any one of those topics is likely to have a few gaps, even though experts in several of these fields have reviewed my chapters and provided me with valuable feedback.

So my ambitious goal has been to create a book that's at least 98 percent correct. The other 2 percent of the information I present might include "facts" that are misleading, oversimplified, out of date, slightly misstated, or downright erroneous. Realistically, 98 percent is not too bad. One of my favorite quotes describes the astronomer and science writer Carl Sagan (who died in 1996) as someone who was "very often right and always interesting." If I can meet a similar standard, to be usually right and almost always interesting, I will have done well. To put it another way, I have high hopes that you will find this book enlightening as well as engaging.

1

The Lungs of the Planet

Ever since I was a child, my friends have noticed a slightly obsessive-compulsive nature to my personality. Just slightly, mind you. For example, I cannot stand to be in the same room as a desk drawer or kitchen drawer that has not been fully closed; I must close it immediately. Simply retreating into the next room will not solve the problem, because I'll still remember that there's an unclosed drawer in the adjoining room. And if I run across a set of data I haven't seen before, I often feel compelled to type that data into a spreadsheet so that I can sort and analyze the data in various ways. Hardly a week goes by in which I haven't created some fascinating new spreadsheet, such as the one that lists all of the species of trees found in a park near my home, along with the botanical family for each species and the typical height of a mature tree. If you ever happen to visit me, I'll be happy to show you my latest spreadsheet.

Another consequence of this personality trait is that whenever I read about science on the internet or in popular publications, I often find myself saying, "Wait! That's not right!" Anyone in the same room with me will soon get an earful about the erroneous material I've just encountered. Quite often, the offending passage is not *completely* wrong; it's just *not quite right*. That was the case the first time I encountered this sentence in a popular science article:

"Rainforests are the lungs of the planet."

What? Really?

Since that first encounter, I have seen several variations on this sentence. Sometimes *rainforests* is replaced by *forests* or *trees*. Sometimes the word *planet* is replaced by *earth* or *world*. But all of these variations convey essentially the same message. The lungs meme has now become quite widespread on the internet and is always expressed as if it were an absolute truth. And by *meme* I don't mean a funny picture or a video, but a concept encapsulated by a short, punchy phrase, such as this metaphor comparing forests to lungs.

On one level, I genuinely appreciate the poetry of this metaphor. On another level, I love the implication that trees and forests have value as living creatures, not just as a source of wood. But an analogy is only as good as the conclusions one draws from it. What conclusions should we draw from this comparison of rainforests to lungs?

What Are Lungs?

You and I each have two lungs, and we use those lungs to breathe. We think of breathing as a two-phase cycle: first we inhale and then we exhale. The air we inhale from the earth's atmosphere is about 78 percent nitrogen, 21 percent oxygen, and 1 percent argon, along with a tiny amount of carbon dioxide (about 4/100 of 1 percent). The air we exhale is obviously different—but not as different as you might think. It is still mostly nitrogen, and still 1 percent argon. The main difference is that *some* of the oxygen has been removed from the air, replaced by carbon dioxide. (The air we exhale also has more water vapor, evaporated from the moist interior of the lungs.)

It is common to say that we breathe in oxygen and breathe out CO_2, but this is far from accurate; we mostly breathe in nitrogen and breathe out nitrogen. Perhaps most surprising, our exhalations contain far more oxygen than carbon dioxide. About a quarter of the oxygen we inhale is absorbed by tiny blood vessels in the lungs, and the rest of the oxygen is exhaled without being used. The same blood vessels that absorb oxygen also give off carbon dioxide, thereby eliminating a waste product from the body. The most important part of breathing is not the inhaling nor the exhaling but the gas exchange—extracting oxygen from the air and getting rid of CO_2.

Many other types of animals besides humans have lungs. On the other hand, certain tiny creatures meet their oxygen needs by absorbing it directly through the skin, without the use of lungs. Even if human skin were optimized for maximum oxygen uptake, a human could never absorb enough oxygen through the skin; we have too much body mass for the amount of skin we have. Lungs solve this problem by presenting an astounding amount of surface area to the air, due to the hundreds of millions of little sacs (called alveoli) inside each lung. Furthermore, these surfaces are always moist, facilitating gas exchange. The muscles in your diaphragm force air in and out of your mouth and nose, inhaling and exhaling, thereby bringing in a fresh batch of outside air every few seconds.

Is It Accurate to Compare Forests to Lungs?

The lungs metaphor implies that forests—especially tropical rainforests—serve as a kind of air exchanger, taking in fouled air and replacing it with clean air, thereby benefiting the whole planet. The underlying idea is that a forest improves the air by removing carbon dioxide and releasing oxygen. On a literal level, this is the opposite of what lungs actually do. Lungs take in fresh air and exhale stale air, partially depleted of oxygen but enriched in carbon dioxide. However, the comparison to lungs is intended as a rough analogy, not a literal fact, so we interpret the metaphor to mean that trees perform the reverse process. Thus a balance is implied between the forests of the world and the animals of the world. In fact, many educational materials contain graphics that illustrate such a balance.

The main strength of this metaphor is its emphasis on gas exchange (the exchange of carbon dioxide with oxygen), which is an important concept. But if a forest has the equivalent of lungs, where are these lungs? The answer is that most of the gas exchange occurs in the leaves. Pores on the lower surface of each leaf (called stomates or stomata) allow gases to move in and out. During the day, carbon dioxide enters through these pores and oxygen escapes. This is consistent with the "reverse lungs" concept. But at night the opposite happens: oxygen enters through the pores and carbon dioxide escapes, a reversal of direction that the lungs metaphor does not explain or even acknowledge. This daily cycle happens because photosynthesis occurs only during the day, but metabolism occurs twenty-four hours a day.

When we think of real lungs, we also think of breathing—alternately inhaling and exhaling. Muscles in the chest first pull air into the lungs and then a few seconds later push the air back out. Do forests "breathe" in a similar manner? Some websites and popular media articles suggest as much, saying that trees "breathe in carbon dioxide and breathe out oxygen." Some go even farther, saying that trees "suck in carbon dioxide," as if trees actually had lungs. Both of these words—*breathe* and *suck*—imply the use of muscles to force air in and out of a lung cavity.

But this is not what happens in a plant. Instead, carbon dioxide and oxygen both slowly diffuse through the open pores in the bottom surface of each leaf, gradually moving from a place of higher concentration to a place where the concentration is lower. When CO_2 is more concentrated in the air outside the leaf than inside, CO_2 slowly enters the pores. When oxygen is more concentrated inside the leaf than outside, oxygen slowly exits the pores. Thus, oxygen and carbon dioxide can pass through a leaf pore *in opposite directions at the same time*— quite different from our usual concept of breathing, in which all the air is forced to go in a single direction at any given moment.

One additional issue with the lungs meme is that it tends to ignore *why* forests produce the opposite results from animal lungs. Rather than simply praising trees for their benefits to us, we should also ask: *Why* do trees remove carbon dioxide from the air? What's in it for the trees? Answering this question—as we will shortly—is

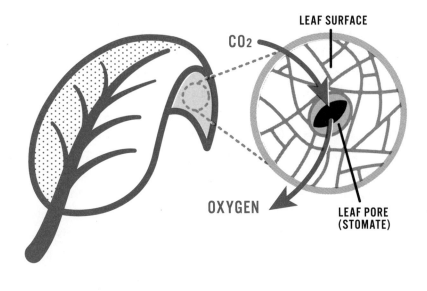

CO2

LEAF SURFACE

OXYGEN

LEAF PORE
(STOMATE)

Gas exchange through a stomate on a leaf

the key to unlocking the underlying science. Unfortunately, a child who has been taught the lungs meme might answer this question by saying, "Because people and animals need oxygen." This confuses a *benefit* with a *cause*. While it's beneficial to us that trees release oxygen and remove CO_2 from the air, trees do it for reasons that have nothing to do with us.

What Is the Point of the Metaphor?

Of course, the reason the lungs metaphor appears so often in the popular media and educational materials is that we (the readers) are supposed to draw an important lesson from it. However, these various sources don't all agree on the point of the lesson.

In some instances, the explicitly stated lesson is that forests produce the air we breathe—and that if we don't stop cutting down trees, we will soon run out of oxygen. ("Forests are the lungs of the earth. If we destroy them, we destroy ourselves!") However, this is a massive exaggeration. Destroying the world's forests would indeed be catastrophic, for many reasons, but it would not result in our suffocating. It is true that all of the free oxygen in our atmosphere was put there by living creatures—a very important point. However, this oxygen has slowly accumulated for several billion years, and it's not going to disappear overnight. Furthermore, trees are not the only organisms that release oxygen into the atmosphere. *All* green plants do so, along with a multitude of microscopic green organisms (algae and cyanobacteria) that live in water and wet places. So while trees are indeed major producers of oxygen, they aren't the sole source.

In contrast, some articles in the media that use the lungs metaphor suggest a far more useful lesson: because trees remove carbon dioxide from the air, they help to offset some of the human-caused increase in atmospheric CO_2. This lesson draws a connection between forests—especially tropical rainforests—and global climate change. If we can slow down or reverse the worldwide reduction in the number of trees, this should help slow the rate of climate change.

So the real point of the lungs meme is not so much the relationship between trees and oxygen as the relationship between trees and carbon dioxide. On the internet and in educational materials, various authors have used a wide range of verbs to summarize this relationship:

- Trees *remove* carbon dioxide from the air.
- Trees *absorb* and *store* carbon dioxide.
- Trees *filter* carbon dioxide from the air.
- Trees *clean* the air.
- Trees *purify* the air.

Each of these phrases represents a slightly different meme intended to encapsulate the relationship. The word *remove* is by far the most accurate of these verbs. Unfortunately, all of these verbs can lead to misconceptions, in part because of several important ideas that these simple memes omit.

Filtering, Absorbing, Storing, Purifying

One popular meme, often associated with the lungs metaphor, is that trees filter the air—equating forests to an air filtration system. The idea is that trees filter out carbon dioxide and other "bad" substances from the air. One advantage of this meme is that it's easy to understand. The better versions of this meme explicitly mention CO_2: trees filter *carbon dioxide* from the air. However, if you take this meme too literally, you might assume that air passes right through the leaves as through a filter, entering from one side of the leaf and exiting the other side, which is not the case.

The filtration meme offers no direct explanation of what happens to the CO_2 that has been removed. This can lead to the misconception that the extracted CO_2 is completely destroyed. On the other hand, if you take the analogy of a filter quite literally, you are more likely to assume that the carbon dioxide accumulates over time in the leaves of plants, which isn't correct either.

What do trees do with the CO_2? A popular concept—similar to the filtration meme but distinct from it—is that trees *absorb* and *store* CO_2. One version of the concept equates a tree to a giant sponge that sops up carbon dioxide from the air, storing it inside the tree. This meme has three important strengths: (1) it's easy to understand, (2) it acknowledges that the carbon dioxide is not magically eliminated, and (3) it subtly implies that the carbon dioxide will return to the atmosphere if the tree is destroyed.

However, this meme also implies that trees serve as storage units for carbon dioxide, which is not correct. Trees *use* carbon dioxide—they don't *store* it. A tree *converts* carbon dioxide into other carbon-based chemical compounds it can use. The great mass of a tree consists primarily of just two things: carbon-based compounds (also called organic compounds) and water. Most of the carbon atoms removed from the air have been incorporated into wood, leaves, or other essential parts of the tree.

Despite the imperfections of this meme, a person who learns it will probably realize that destroying a forest has *two* negative effects connected to carbon dioxide. First, there are fewer trees to remove carbon dioxide from the air. And second, destroying a forest tends to release a lot of carbon dioxide into the atmosphere in a short period of time.

The relationship between trees and carbon dioxide is sometimes expressed in the popular media with the verbs *clean* and *purify*, as in "trees clean the air" or "forests purify the air." It is true that trees can reduce the concentration of certain harmful pollutants in the air, such as ozone, sulfur dioxide, nitrogen dioxide, and particulates (soot and dust). But applying these verbs to carbon dioxide just muddies the water.

The verb *purify* is especially misleading, because it implies that carbon dioxide in the air is an impurity. Carbon dioxide occurs naturally in the air, and green plants depend on it to survive. Therefore, our entire food supply depends (directly or indirectly) on the presence of CO_2 in the air. The real issue is that when the amount of CO_2 in the atmosphere changes, either increasing or decreasing, it causes climates all around the world to change, which is disruptive to both human societies and natural ecosystems. It's actually quite good that the atmosphere contains CO_2, but it's bad that human activity is causing the concentration of CO_2 in the air to increase so rapidly.

The verb *purify* is misleading in other ways. First, it greatly exaggerates the results. Trees can *reduce* the level of pollutants in the air, but they fall far short of actually purifying it. Second, trees also pump large quantities of material *into* the air. Many species of trees are wind pollinated, including oaks, maples, birches, hickories, pines, junipers, and poplars. A single mature tree can release hundreds of millions of pollen grains into the air each year, to the dismay of people who suffer from spring allergies (as I do). Many trees also scent the air by releasing odoriferous chemicals. Of course, we usually perceive these odors as pleasant, such as the smell of pine, juniper, or eucalyptus (or my personal favorite, California bay laurel). In effect, the chemicals released into the air serve as nature's air freshener. (Perhaps this explains why people hang tree-shaped air fresheners in their cars!) Furthermore, insect-pollinated trees, such as the southern magnolia, can release wonderful scents when the flowers are in bloom, thereby alerting pollinators that dinner is served. But none of this counts as purifying the air. When we smell the fresh scent of a forest, it's not because the air has been purified but because of the natural chemicals that have been released into the air.

Why Do Trees Remove CO₂ from the Air?

If the key lesson of the lungs metaphor is that trees remove CO_2 from the air, our lesson isn't complete until we understand *why* trees do it. The answer, in a word, is *photosynthesis*. As we were all taught in school, green plants use photosynthesis to capture the energy of sunlight. In this abbreviated form, the concept seems to be unrelated to our discussion about trees and carbon dioxide. However, a slightly longer version spells out the connection: green plants use the energy of sunlight to convert carbon dioxide and water into sugar, releasing oxygen as a by-product.

Unlike the memes discussed earlier, the concept of photosynthesis provides a *reason* that plants remove carbon dioxide from the air: to produce sugar. It also explains what happens to the carbon: it becomes part of the sugar molecule ($C_6H_{12}O_6$). This explanation also implies how green plants benefit from the process: they can use the sugar.

The diagram on the following page indicates the specific molecules involved in photosynthesis, but to produce a balanced equation you would have to mention the *quantities* of each molecule: *six* molecules of CO_2 and *six* molecules of water combine to form *one* molecule of glucose plus *six* molecules of oxygen.

Note the detail that oxygen is given off as a waste product of photosynthesis. Carbon dioxide and water contain more oxygen atoms than are needed to make sugar, so the excess oxygen is released as a gas. That's the reason green plants give off oxygen—not because animals and humans need it. In fact, when the earliest photosynthetic organisms began to pump oxygen into the atmosphere three billion

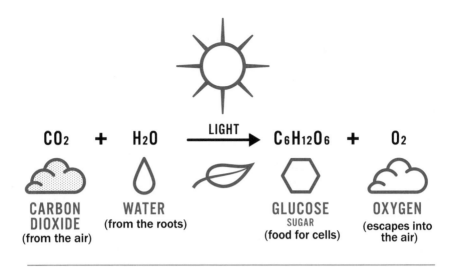

$$CO_2 + H_2O \xrightarrow{\text{LIGHT}} C_6H_{12}O_6 + O_2$$

CARBON DIOXIDE (from the air) + **WATER** (from the roots) → **GLUCOSE** SUGAR (food for cells) + **OXYGEN** (escapes into the air)

The process of photosynthesis

years ago, the gas poisoned much of the existing life on Earth, killing it off but paving the way for the later evolution of oxygen-dependent creatures. (This episode may sound quite sad, but that ill-fated early life was mostly anaerobic bacteria of various kinds.)

This simple model of photosynthesis—using sunlight to convert CO_2 and water into sugar—provides a great foundation for understanding the relationship between trees and carbon dioxide. However, this model is incomplete because it fails to explain what happens to all that sugar. The simplest such explanation (although still incomplete) is that the sugar produced by photosynthesis serves as food for the plant. This is a crucial concept. Every living cell needs energy to survive, and for most plant and animal cells, this energy is delivered as sugar. The sugar produced in the leaves of a plant must be transported to all the living cells in the plant, including the roots.

Once you fully grasp these two ideas—that every plant cell needs food in the form of sugar and that a living plant must move sugar to where it's needed—it makes perfect sense that most land-based green plants have an internal water-based transport system. In fact, two distinct transport systems are at work. One system moves sugar water down from the leaves to the roots, and the other system moves mineral water up from the roots to the leaves.

Why do plant cells need energy? Cells use the chemical energy of sugar to drive the normal metabolic processes that keep the plant alive. When the cells use this energy, the sugar reverts to carbon dioxide and water—and oxygen is consumed in the process. The upshot is that every cell in a plant constantly consumes oxygen and gives off carbon dioxide, just as animal cells do. However, when the sun is shining, the chloroplasts in the leaves and other green surfaces do just the opposite, and they do it at a much faster rate. Thus, during the day, green plants are net consumers of carbon dioxide and net producers of oxygen. But at night, when photosynthesis shuts down, it's just the opposite.

In short, to truly make sense of the concept of photosynthesis, one must remember the following three details:

1. Plants use the energy of sunlight to convert carbon dioxide into sugar.
2. The creation of sugar molecules is a way of *storing* the energy of sunlight.
3. Sugar is the principal source of energy for living cells.

But even if you remember these three details, the story is not complete—because sugar molecules provide a second benefit that is just as important as storing energy.

Making Useful Stuff from Sugar

What is that second benefit? you may be asking. The concept missing from the preceding discussion is that much of the sugar produced by green plants is *not* used to provide energy to the cells of the plant. Instead, the sugar is converted into other organic compounds that are useful to the plant. A surprisingly wide range of compounds is produced, including starches, fats, proteins, and many other classes of molecules. Some of these compounds, such as starches and fats, require nothing more than the atoms already present in sugar—carbon, hydrogen, and oxygen. But some compounds (such as proteins) require additional atoms (such as nitrogen) that arrive via the mineral water sent up from the roots. These various molecules serve many different purposes in the life of the plant.

In most plants, a high percentage of the sugar that is created is converted into cellulose—or in the case of woody plants, cellulose and lignin. These are the structural materials that give a plant its shape and allow it to stand upright. (Lignin, which is much stiffer than cellulose, is the compound that makes woody plants "woody.") Humans cannot digest cellulose or lignin, so we tend to eat the parts of plants where the digestible compounds—such as sugars, starches, fats, and proteins—have been concentrated.

All that useful stuff plants make tends to build up over time. This brings us to the concept of biomass. Contrary to what the word sounds like, *biomass* is not a religious ceremony for science teachers. Instead, biomass is any material that consists either of living tissue—plant or animal—or matter that was once living. In a forest ecosystem, most of the biomass consists of living trees or dead remnants of trees, such as the leaf litter on the forest floor. Some of the biomass is underground, including tree roots, fungus, other microorganisms, and the myriad little critters that live in the soil.

One component of biomass is water, embedded in living or dead tissue. But the rest of the biomass consists almost entirely of energy-rich carbon-based compounds. For that reason, dried biomass is flammable and can be used as fuel. The most obvious example is firewood, but any dried plant material tends to burn easily. This fact reveals a key detail: cellulose and lignin contain a lot of stored chemical energy. In fact, all the carbon-based compounds in a plant are high-energy, and this energy can be traced back to sugar created by photosynthesis.

The only organisms that can convert CO_2 to sugar are green plants and green microorganisms (algae and cyanobacteria), both of which contain chlorophyll. These are the only organisms that can create new biomass. (One minor exception is organisms that use inorganic chemical energy instead of sunlight to create new biomass, such as the bacteria around deep-sea hydrothermal vents.) Animals cannot create new biomass, but they can convert part of the biomass they eat into other kinds of tissue. However, doing so always results in a net loss of biomass. In other words, when an animal eats biomass (plant or animal tissue), only a small part of that biomass is incorporated into the body of the animal as muscle or other tissue. A larger part of that biomass is simply metabolized for its energy. And a far larger part of the eaten biomass is wasted, especially if the animal is incapable of digesting cellulose. The key point here is that in a typical terrestrial ecosystem, such as a forest or grassland, all of the biomass in the system is originally created by plants. (In an aquatic ecosystem, algae and cyanobacteria—which are also photosynthetic organisms—often fill the role instead.)

When discussing the biomass of an ecosystem, it's helpful to consider how dense the biomass is. This can be expressed, for example, as tons of biomass per acre or metric tons per hectare. Not surprisingly, forests (especially tropical forests) tend to have very high values because so much biomass is locked up in woody tree trunks, branches, and roots.

Did I Hear Someone Say "Carbon Sink"?

If the term *carbon sink* makes you think of a high-tech bathroom fixture, I'm about to open your eyes to a completely new meaning. A carbon sink is anything that absorbs large amounts of carbon dioxide from the atmosphere, retaining the carbon in one form or another. Because forests have such high biomass density and all biomass consists of carbon-rich compounds that originated as atmospheric CO_2, forests can be viewed as a major carbon sink.

However, a carbon sink doesn't always remain a sink; the flow of carbon atoms can easily reverse direction. The biomass of a forest reverts to CO_2 again whenever any of the following happens:

- Sugars are metabolized by plant or animal cells in order to access the stored energy.

- Dead biomass, such as fallen leaves or downed trees, decomposes into simpler compounds. (Decomposer organisms play a key role, utilizing some of the stored energy while breaking down the organic compounds.)
- Fire races through a forest, burning the dead forest litter—and in the case of a crown fire, also consuming parts of living trees.

In a typical forest, far more carbon is captured than released, although the amount varies according to the type of forest, the age of the forest, and other factors. Recent studies have explored this issue in detail, examining a wide range of forests around the world. They show that a typical forest continues to gain biomass until the forest is about eight hundred years old, after which the quantity of biomass remains at a steady state and the forest becomes carbon neutral.

This result may seem counterintuitive, especially if you picture a forest as reaching maturity in less than a century. But consider the most massive trees in any typical forest, such as the largest species of oaks in many temperate forests. These trees can live for hundreds of years, gaining biomass in their trunks every year (because the diameter continues to increase as long as the tree is alive). Furthermore, after the forest finally reaches a steady state in its aboveground biomass, the soil carbon continues to increase for a while. The upshot is that a typical forest continues to capture additional carbon for about eight hundred years. Most forests in the world are far younger than that, in part because humans have cut them down at one time or another.

At the other end of the age spectrum, freshly harvested forest land—even if replanted with young trees—continues to lose carbon dioxide to the air for about fifteen years before finally becoming a carbon sink again. That is due to the decomposition of all the dead tree parts left behind—branches, leaves, stumps, and roots—and the loss of some of the existing soil carbon. But for a typical forest in the age range of fifteen to eight hundred years, the amount of stored carbon continues to increase over time.

Because trees can be very large, it seems intuitive that a forest would store more carbon per acre than any other type of ecosystem.

But is that really true? If you consider only the aboveground storage of carbon, the tropical rainforests of the world are the clear winners. Forests in temperate climates also store a lot of carbon, but less than tropical forests.

However, if you also consider the organic carbon stored in *soils*, the picture becomes more complicated. In the extensive peatlands of the world, the density of carbon storage can be as great as in tropical forests. However, much of this carbon is stored in a thick blanket of peaty soil, not in living vegetation. The acidic, waterlogged soils prevent fallen organic matter from decomposing, so it builds up over a long period of time. Peatlands are especially common in the far north—Canada, Russia, and Alaska—but the tropics also contain significant areas of peatland. The destruction of forested peatlands in Indonesia and Malaysia to make way for palm oil plantations is particularly significant because the combination of forest with peaty soil is especially carbon rich.

Destroying peat bogs is as bad as destroying tropical forests when viewed through the lens of preserving our major carbon sinks. Peat bogs are easily destroyed by draining away the water, which exposes the soil to air, allowing the organic matter to decompose. However, peatlands are not the only ecosystem with high levels of organic carbon in the soil. Mangrove swamps tend to have very high levels of soil carbon, and grasslands tend to have fairly high levels. Worldwide, more organic carbon is found in the top meter of soil than in all the aboveground biomass.

The ocean is also a major carbon sink, because carbon dioxide is soluble in water and easily passes between the atmosphere and the ocean. In fact, the ocean contains far more carbon dioxide than the atmosphere. Thus, the three major carbon sinks of the world are vegetation, soil, and oceans, and each is capable of returning carbon dioxide to the atmosphere, depending on current conditions.

To round out this picture, it is also helpful to think about the *former* carbon sinks of the world, now locked away deep in the earth. The two former carbon sinks are fossil fuel reserves and limestone in the earth's crust.

Our reserves of fossil fuel—oil, gas, and coal—are the remnants of ancient swamps in which large amounts of plant material accumulated without decomposing. This organic matter eventually became buried under soil thousands of feet deep. The combination of heat, pressure, and time converted the buried soil into sedimentary rock, and the embedded organic material into petroleum, natural gas, and coal. These fossil fuels have been locked away for hundreds of millions of years, but now humans actively seek out these reserves to burn as fuel, which returns the carbon dioxide to the air.

The vast amounts of limestone in the earth's crust are a result of the presence of carbon dioxide in the oceans. CO_2 combines with water to form carbonate, which remains dissolved in the water. Many forms of sea life extract carbonate to produce shells, reefs, and other hard structures. Additional carbonate interacts with calcium that has weathered from continental rocks and washed into the ocean. Both of these processes result in a steady rain of calcium carbonate settling to the bottom of the ocean, forming thick layers of marl that eventually become limestone and related rocks. When limestone is processed to create cement, some of the carbon dioxide returns to the air.

The Big Picture

But what about the original question? Are rainforests the lungs of the planet? If we turn this into a true/false question, the best answer is false. In no practical sense can we accurately say they are the lungs of the planet. Every molecule of CO_2 captured from the atmosphere and converted into biomass results in the release of one molecule of oxygen. The atmosphere has five hundred times as many oxygen molecules as CO_2 molecules, so if plants somehow converted every single molecule of atmospheric carbon dioxide into biomass, this would have no appreciable effect on the concentration of oxygen in the air. (However, if this imaginary scenario actually happened, CO_2 would seep from the ocean into the air, restoring part of the missing CO_2.)

Likewise, humans and animals don't use enough oxygen to significantly affect the amount of it in the atmosphere.

On the flip side, while it is very helpful to preserve forests and to plant new trees—for many reasons, including our fight against global warming—forests will never generate enough biomass to completely compensate for all the fossil fuel CO_2 we have poured (and continue to pour) into the atmosphere. In other words, trees alone cannot reduce the levels of atmospheric carbon dioxide to their preindustrial levels.

Now that we have thoroughly tugged at the loose threads in the lungs metaphor, the sweater has become seriously unraveled. But perhaps this threadbare garment still has some use. Here are the five principal concepts to take away from this discussion:

1. The lungs metaphor is all about gas exchange—oxygen and carbon dioxide—and how forests do it in the opposite direction from animals. But the most valuable idea in this metaphor is that trees remove carbon dioxide from the air.

2. Trees don't remove CO_2 because of their deep compassion for people. Instead, they need the CO_2 to make sugar, thereby capturing the energy of sunlight. This process, called photosynthesis, gives off oxygen as a waste product.

3. Sugar is the principal source of energy for living cells—in both plants and animals. When a cell uses this energy, the sugar reverts to CO_2 and water, consuming oxygen in the process.

4. Plants convert some of the sugar they make into a wide range of other essential organic compounds, providing structure and shape, as well as food for the various animals in the ecosystem.

5. The forests of the world are a major carbon sink, storing additional biomass as they grow. Destroying a forest releases most of the carbon back into the atmosphere.

Taken together, these five concepts provide a detailed and nuanced picture of the relationship between trees and carbon dioxide.

Of course, the forests of the world provide far more benefits than just capturing carbon—and the wholesale destruction of forests does far more harm than just releasing carbon dioxide into the atmosphere. But with the current emphasis on trees as part of the solution for fighting the rising levels of atmospheric CO_2, it is helpful to have a good understanding of the underlying scientific concepts.

I'm happy that I was able to wrap up this chapter with a nice little list of five key points. I often think lists are even more interesting than spreadsheets. I've got lists all over the place. I even keep a list of the lists I intend to make. However, if your personal preference is to see a spreadsheet rather than a list, just let me know and I'll see what I can do.

2

No Gravity in Space

I was busy arranging my coffee mugs the other day when I heard a shocking statement on the radio. I was actually paying more attention to the mugs than the radio—lining them up on a kitchen shelf, equidistant from each other, with the handles on the right and the principal image or text facing forward. I used a ruler (marked in millimeters) to ensure I was getting it right. But then I heard a voice on the radio mention that the International Space Station is a great place to conduct certain scientific experiments because there is no gravity in space. I was so stunned that I nearly dropped my favorite mug, the one that depicts the periodic table of elements. *No gravity in space?* How could they say such a thing on the radio? I promptly jotted down a note to remind myself to address this unfortunate misconception.

It's true that astronauts in Earth orbit experience weightlessness—and that this effect is sometimes misleadingly called zero gravity—but weightlessness does not actually indicate a lack of gravity. In fact, quite a lot of gravity is tugging on the space station. The space station orbits in a continuous loop around Earth, completing each orbit in about ninety-three minutes. This nearly circular path is the result of Earth's gravity. Without Earth's gravitational attraction, the space station would fly off in a straight line, rapidly leaving the vicinity of Earth. The same thing applies to the moon, whose orbit is nearly a thousand times as far from Earth as that of the space station. It, too, would fly off if not for the effect of Earth's gravity.

Of course, this leads to an apparent contradiction. If Earth's gravity is strong enough to bend the path of the space station into a circle, how can the astronauts—and everything else inside the space station—experience weightlessness? If the space station is subject to Earth's pull, shouldn't the people inside feel the effects of gravity too?

Einstein's Elevator

A helpful place to start is with a famous thought experiment conducted by Einstein a century ago. Imagine yourself standing in a windowless elevator. You have not yet pushed a button, and therefore the elevator is not yet moving. You are aware of Earth's gravity because your legs support your weight. Because of gravity, you have no doubt as to which way is up and which way is down. Now imagine that this same elevator, with you inside, is magically transported far into space, far from any source of gravity such as a planet or a star. If the elevator is not moving (or more accurately, not accelerating), you will float around inside the elevator in a state of weightlessness. The only clue as to which way is up is the interior design of the elevator.

Now imagine that this elevator, still out in deep space, begins to accelerate in the direction of up, consistent with the interior of the elevator. Instead of experiencing weightlessness, you will again be standing on the floor of the elevator. If we assume that the elevator accelerates at a constant rate, you will feel a constant downward force that is indistinguishable from gravity. Depending on the rate at which the elevator accelerates, this downward force could exactly match the familiar effect of Earth's gravity. Standing in that elevator accelerating through space, you will not be able to distinguish the experience from an unmoving elevator back on Earth. Even if you play a game of basketball or ping-pong inside the elevator, you will not be able to distinguish between accelerating through space and sitting still on Earth. This thought experiment told Einstein that a gravitational field is indistinguishable from acceleration. In fact, he concluded that the two are equivalent, and this insight played a key role in his formulation of the theory of general relativity.

Let's continue the exercise. Suppose you enter an elevator at the top floor of a very tall building. Imagine that this particular elevator is capable of accelerating downward at a rate that exactly balances the effect of gravity. You press the button to travel to the ground floor, and for a few seconds you experience complete weightlessness. (However, before reaching the bottom, the elevator must sharply decelerate, resulting in several seconds when your weight is much greater than its normal value.) Even though you experience a period of weightlessness, you remain under the influence of Earth's gravity the entire time. Your weightlessness is due not to a lack of gravity but to a balance between two effects. Without the effect of gravity, the downward acceleration of the elevator would plaster you against the ceiling. But the downward force due to gravitational acceleration exactly balances it, so you float weightlessly within the elevator.

The training program for future astronauts puts this idea to practical use. A special jet airplane (the "vomit comet") produces the effect of weightlessness. After climbing high into the sky, the jet dives toward the earth, exactly matching the rate of gravitational acceleration. In other words, the people inside are falling toward the earth due to gravity, but the downward motion of the jet matches the increasing speed at which they are falling. With each dive, the passengers on this jet experience twenty to thirty seconds of weightlessness, floating around a padded cabin within the plane. During this entire period of weightlessness, Earth's gravity never disappears; it is simply masked by the downward acceleration of the jet.

Passengers aboard the space station—or in any other space capsule orbiting Earth—experience weightlessness for almost exactly the same reason as on that diving jet. However, before we can make sense of this idea, we have to explore another related concept.

Newton's Apple

When I was a kid, I was told that Newton discovered gravity when an apple fell on his head. It's a cute story, but no evidence exists that a falling apple ever collided with Newton's noggin. Furthermore, this story completely misstates the essence of Newton's discovery. People were already quite familiar with gravity long before Newton. It was an everyday experience that if you drop an object, it falls to the ground. Newton's genius was to ask questions about gravity that most people failed to ask, and then to discover answers to those questions.

Newton had indeed grown up around apple trees, and he had seen apples fall. On a windless day, an apple falls in a straight line directly toward the ground—in other words, perpendicular to it. But on a windy day, a falling apple follows a curved path to reach the ground. For a mind like Newton's, this observation would lead to lots of questions, such as: Why did the path of the apple curve? What is the exact shape of this curve? What would happen if the wind were much stronger?

This last question is especially fascinating. In a stronger wind, the apple would fall farther from the tree. Increase the wind even more, and the apple would fall farther still, tracing a more elongated curve. However, the surface of the earth also forms a curve. Imagine a wind so strong that the curve of the apple's path exactly matches the curvature of the earth. The apple would just keep circling the earth forever, never getting any closer to the ground. Newton knew that this would never actually happen to an apple, but he realized that this idea could apply to the moon in its orbit around the earth, or to the planets in their orbits around the sun.

Unfortunately, the apple analogy has a serious weakness: it's unclear whether the gust of wind just gives the apple an initial shove or if the wind continues to push the apple during its fall. So when Newton wrote up his laws of motion, he used the example of a cannonball instead. Newton imagined a cannon sitting atop a very high mountain. The cannon fires horizontally, shooting the cannonball parallel to the surface of Earth. (Note that the cannon does not influence

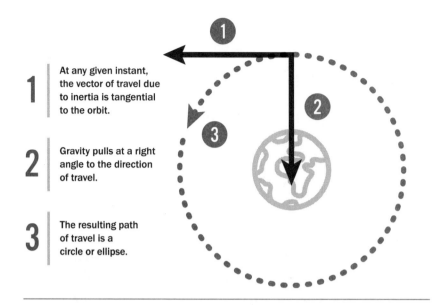

1. At any given instant, the vector of travel due to inertia is tangential to the orbit.

2. Gravity pulls at a right angle to the direction of travel.

3. The resulting path of travel is a circle or ellipse.

How the combined effect of inertia and gravity can produce a circular orbit

the cannonball after it exits the cannon's bore.) Because of gravity, the cannonball eventually falls to the ground, tracing a curved path. But if the cannon were to eject the cannonball with sufficient speed, the arc of the falling cannonball would exactly match the curvature of the earth. As long as the cannonball never slowed down, it would keep circling the earth. Of course, air friction would indeed cause the cannonball to slow down, but air friction does not affect the motion of the moon around the earth, or the earth around the sun.

Newton also realized that the trick for describing the curve of a falling object is to consider the horizontal motion separately from the vertical motion. A gust of wind gives the apple its horizontal motion, and the firing of the cannon gives the cannonball its horizontal motion. But in both cases, gravity gives the object its vertical motion. The horizontal speed of the falling apple or flying cannonball is essentially constant, while the vertical speed continues to increase due to gravitational acceleration. This combination of motions produces a curved path.

This same idea—that a combination of horizontal velocity and vertical acceleration produces a curved path—also applies to any object in orbit around the earth. Take the example of the space station, which maintains a roughly constant distance from Earth and an essentially constant speed in its orbit around Earth. At that elevation, where the friction of Earth's atmosphere is quite small, the horizontal component of motion is entirely due to inertia. *Inertia* is the tendency of an object to keep traveling in a straight line at a constant speed in the absence of external forces. The space station does not rely on rocket engines to keep it moving forward—and yet it keeps moving forward. If not for the gravity of the earth, the path of the space station would indeed be a straight line. But the gravity of the earth keeps bending its path toward the earth, just enough to match the curvature of the earth. The space station is always falling toward the earth, but it never gets any closer to the earth. The falling toward the earth is exactly balanced by the inertial tendency to travel away from the earth on a straight, tangential path.

This is why an astronaut aboard the space station experiences weightlessness. Everything inside the space station is falling toward Earth at exactly the same rate as the space station itself. Earth's gravity is still present, acting on the space station and everything inside it. But the net result is weightlessness, just as on the "vomit comet" training jet.

You Can't Orbit at Just Any Speed

Maintaining Earth orbit requires a balance between the motion of falling (due to gravity) and the motion of flying away (due to inertia). It follows that speed is a crucial factor. The space station orbits Earth at 17,100 miles per hour. If it orbited any more slowly, the two effects would no longer be in balance, and the space station would gradually lose elevation, eventually crashing into Earth. If it orbited much faster, the space station would gain elevation and eventually fly off into space.

The pull of Earth's gravity diminishes as you move away from the earth. This means that the farther you get from Earth, the smaller the amount of gravitational acceleration. Therefore the proper speed for maintaining orbit depends on the height of the orbit. The moon, at 239,000 miles from the earth, needs to travel at 2,200 miles per hour to maintain its orbit—only one-fifth the orbital speed of the space

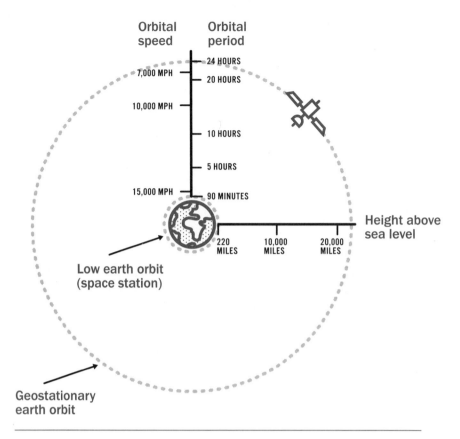

Orbital speed	Orbital period		
	24 HOURS		
7,000 MPH	20 HOURS		
10,000 MPH			
	10 HOURS		
	5 HOURS		
15,000 MPH	90 MINUTES		
	220 MILES	10,000 MILES	20,000 MILES

Height above sea level

Low earth orbit (space station)

Geostationary earth orbit

Comparing low earth orbit to a geostationary orbit

station (which orbits approximately 240 miles above the earth). It takes the moon about twenty-nine days to circle the earth, compared to ninety-three minutes for the space station. If the moon were traveling at the same speed as the space station, it would complete its orbit around the earth in just six days, passing through all the lunar phases from full moon to new moon and back to full moon. But if the moon really did travel as fast as the space station, it would be moving too fast to stay in Earth's orbit, and it would fly off.

Now imagine a satellite orbiting the earth at a greater distance than that of the space station but not as far away as the moon. The time to complete one orbit would have to be more than ninety-three minutes but less than twenty-nine days. The actual amount of time would depend on the height of the orbit. Imagine that this satellite is just the right distance to require twenty-four hours to complete one orbit. If the orbital path were aligned with the spin of the earth (that is, traveling directly to the east and directly above the equator), the satellite would appear to hover at single point above the earth. From the standpoint of someone on the ground, the satellite would appear not to be moving at all, and thus this is called a geostationary orbit. It can be very helpful to park certain kinds of satellites (such as communications satellites) in such an orbit, which is about a hundred times as far from Earth as the space station and other low-orbit satellites but only one-tenth as far away as the moon. Geostationary satellites travel around the earth at 6,876 miles per hour—much slower than the space station but much faster than the moon.

The Tough Issue I've Avoided:
What Is Gravity?

I've come to the point where I must acknowledge the giant elephant in the room. Newton and Einstein didn't exactly see eye to eye about the nature of gravity. Newton's model of gravity conforms rather well to most of our intuitive perceptions regarding space, time, and motion. Einstein's explanation of gravity is more technically correct than Newton's, but it is quite different, and it clashes horribly with our intuition.

Newton thought of gravity as a force, and based on this concept, he developed a set of laws and formulas that precisely describe the motion of objects in the solar system. Newton's formulas are still widely used by engineers and physicists today for all sorts of purposes. However, there are some phenomena that Newtonian physics cannot explain. Einstein looked at gravity in a different way, as a bending of space-time, and this approach allowed him to develop a different set of laws and formulas that are capable of describing these other cases. As a result, we continue to use Newtonian physics for the many situations where it provides accurate results, but we switch to the more complex Einsteinian physics when there are "relativistic effects" that must be accounted for. For example, Newton's laws work fine for putting a satellite into orbit, but to calculate our location based on communication with GPS satellites, we have to use Einstein's formulas.

In Newtonian physics, we describe gravity as an attraction between two masses—which implies that gravity acts as a force. The bigger the mass, the greater the gravitational attraction it exerts on other masses. But distance is also a factor. Two objects that are close together will exert a greater force on each other than those same two objects when they are farther apart.

As he explained in his theories of relativity, Einstein saw gravity as a warping of space-time in the vicinity of any mass. A large mass warps space-time more than a small mass does. The amount of warping varies with the distance from the mass, decreasing as the distance increases. This concept, even though it has been proven correct, is not intuitive at all. We can easily perceive three-dimensional space, and we can easily perceive time—both of which are key elements in Newtonian physics. It is far more difficult to grasp the idea of merging those concepts into a seamless four-dimensional space-time.

What I've said so far may already seem weird, but it gets even weirder. In Einstein's model, you could say that our perception of gravity is simply an illusion caused by our inability to perceive space-time. Any object in orbit around another object is actually traveling through space-time in a straight line and at a constant velocity—in other words, it is "inertial" with respect to space-time. This differs sharply from our concept of inertia in Newtonian physics. And remember when we discussed Einstein's realization that gravity and acceleration are equivalent? This means that whenever you can feel the effect of gravity, you are actually accelerating through space-time. Thus, if you are standing on the ground, you are accelerating—but if you are feeling weightless (as in the plummeting "vomit comet"), you are *not* accelerating. Again, this sharply contrasts with our concept of acceleration in Newtonian physics. It all boils down to whether we place our frame of reference in three-dimensional space (as Newton did) or in space-time (as Einstein did).

But if Newtonian physics is good enough to put the space station into orbit around the earth, it should be good enough for our discussion of gravity. It certainly makes the explanation easier to follow. Everyday physics is full of examples in which we treat gravity as a force. Furthermore, gravity is considered to be one of the four fundamental forces of nature (along with the strong nuclear force, the electromagnetic force, and the weak nuclear force), although we can dodge the issue by saying "the four fundamental interactions" instead. For the purpose of this discussion, let's continue to think of gravity in the same way Newton did—as a force or an attraction between two bodies. Just remember that this is not strictly correct. But true or not, it does give us a very useful model that produces excellent results, which is sometimes all that really matters.

The Long Reach of Gravity

How far can gravity reach? Gravity causes the moon to circle the earth, but gravity also causes the earth to circle the sun—at a distance of 93 million miles. The planet Neptune also circles the sun—at a distance of 2.8 *billion* miles, which is thirty times as far away as the earth. An astronomical unit (AU) is defined as the average distance between the earth and the sun, so we could say that Neptune's orbit has a radius of 30 AU. Other objects in the solar system orbit the sun at distances much greater than Neptune. Beyond Neptune lies the Kuiper Belt, and beyond that is the Oort Cloud, the apparent source of most comets. Some objects in the Oort Cloud are more than 100,000 AU from the sun—which is more than a light-year (the distance light travels in a year)—and yet these objects remain in orbit around the sun due to gravity.

Gravity can have an effect at much greater distances than the Oort Cloud. You have probably seen photos of spiral galaxies, looking like gigantic pinwheels. Other galaxies have a more globular appearance. But in either case, the billions of stars in a galaxy cluster together because of gravity, resulting in that pinwheel or globular shape. The upshot is that gravity can operate at distances of millions of light-years or more. Because a typical galaxy contains billions of stars, the total mass of a galaxy is enormous. The greater the mass involved, the greater the distance at which the resulting gravity can have a significant effect. If you consider the tremendous amount of gravity needed to shape a galaxy, it makes sense that in most galaxies, the center of the galaxy contains a black hole. A black hole is simply an immense mass that produces so much gravity that even light cannot escape, making it look like a great void in space.

On the other hand, the gravitational effect of a mass decreases rapidly as you move away from the mass. A simple formula called the inverse square law describes how gravity decreases as the distance increases: $1/x^2$. To use this formula, you first compare two distances, dividing one by the other, then use the result to replace the x to see how gravity is affected. For example, if planet B is twice as far from the sun as planet A, dividing one distance by the other gives you a value of 2. When you plug 2 into the formula, you get $1/2^2$, which is 1/4. This means that if the two planets happen to have exactly the same mass, the sun's pull on planet B is only 1/4 as strong as on planet A. If planet B is instead ten times as far from the sun as planet A, plugging 10 into the formula gives you 1/100, which means that the sun's gravitational attraction is only 1/100 as great. The inverse square law also applies to the intensity of light. At the distance of Neptune, thirty times farther from the sun than Earth, plugging 30 into the formula tells us that the sunlight is only 1/900 as strong as on Earth. Taking vacation photos at noon on Neptune would be like shooting photos in a dim room on Earth. Such low light presents a challenge when photographing the planet, whether from Earth or from a spacecraft like *Voyager 2*.

With regard to gravity, the inverse square law is based on the distance not from the surface of a massive object but from the center of its mass. This distinction is unimportant when discussing the distances between the planets or stars, but it is quite important when discussing satellites in orbit around the earth. Although the space station orbits at a distance of only 220 miles above the surface of the earth, this orbit is almost 4,200 miles from the center of the earth. By this measure, a geostationary satellite is not one hundred times as far from the earth as the space station but only about six times as far.

More to the point, a person standing on the surface of the earth is nearly 4,000 miles from the center of the earth—not much different from the astronauts in the space station. The pull of earth's gravity is diminished by only 10 percent at the elevation of the space station. If the space station could just hold still for a moment, freezing in place instead of orbiting, the astronauts on board would weigh 90 percent as much as they do on Earth. Likewise, if the space station began to travel in a straight line instead of a circle but still at a constant speed, everyone on board would regain 90 percent of their normal weight—at first. By traveling in a straight line, the space station would move away from Earth, and all of the astronauts would gradually become lighter.

Gravitational Acceleration

Imagine you're holding an orange 5 feet above the floor, and then you release it. The orange quickly moves toward the floor. However, the orange does not move at a constant speed. Instead it accelerates as it moves toward the floor. While it's obvious that the orange transitions from not moving to moving, which implies acceleration, it's less obvious that the orange continues to accelerate as it falls.

If you switch from an orange to a hard rubber ball, it becomes a bit easier to see this acceleration. If you drop the rubber ball onto a hard floor, it will bounce back up nearly as high as the point from which you released it. Then the ball will fall back to the floor and bounce back up again. As this motion is repeated over and over, each bounce becomes a bit smaller than the previous one. If you carefully watch this motion, it becomes obvious that in its upward journey, the ball moves fastest when it first leaves the floor and slowest just before it changes direction in the air. Study a bit longer, and it becomes obvious that the downward journey is the mirror image of the upward journey—slowest at the top and fastest just before hitting the floor. In other words, the ball constantly changes speed as it bounces up and down.

The reason for this change of speed is that the force of gravity never stops tugging on that ball. During its downward journey, the constant tugging of gravity causes the ball to move faster and faster. When the ball hits the floor, an elastic collision causes the ball to reverse direction and fly upward, initially traveling at nearly the same speed as when it hit the floor. But now the constant tug of gravity causes the ball to move more and more slowly, until it finally reverses direction again, and the process then repeats.

The fact that gravity causes a falling object to accelerate allows us to talk about gravitational acceleration—the rate of acceleration caused by gravity. Our intuition may tell us that a 5-pound weight will fall faster than a 1-pound weight, but in fact both fall equally fast. If you hold up two heavy objects of different weights—one in each hand—and release them simultaneously, they will hit the ground simultaneously. This was the whole point of Galileo's famous experiment of dropping two objects simultaneously from the Leaning Tower of Pisa. (We don't know if Galileo actually dropped weights from this particular building, but he certainly conducted similar experiments in other locations.) Of course, if you drop a feather or a parachute, it will fall more slowly than a lead weight, but that is because of the large amount of air resistance relative to the weight of the object.

If you limit yourself to objects that have very little air resistance, you will find that all objects fall at the rate of approximately 32 feet per second per second (in other words, 32 feet per second squared), abbreviated as 32 ft/sec^2. If I had just said "32 feet per second," my phrase would have suggested a constant velocity. But when I say "per second" twice (or else "per second squared"), my meaning is quite different, even if a bit obscure. The phrase actually implies a constantly increasing velocity—in other words, acceleration. At the end of one second, a falling object achieves the speed of 32 feet per second. But after every additional second, the speed increases by another 32 feet per second. Therefore, after two seconds, the object falls at 64 feet per second. After three seconds, the object falls at 96 feet per second. In a vacuum, the speed would continue to increase until the object hit the earth. But in the earth's atmosphere, air resistance eventually prevents any more acceleration (because the force of the resistance exactly matches the force of gravity) and the object reaches terminal velocity.

Of course, most of us don't normally discuss speeds in terms of feet per second. Instead, we usually talk about miles per hour. A speed of 32 feet per second is approximately 22 miles per hour. Expressed this way, a dropped object should reach a speed of 22 miles per hour after one second, 44 miles per hour after two seconds, and 66 miles per hour after three seconds. For a skydiver, terminal velocity is about 120 mph—so it doesn't take very long to reach this speed. However, it takes longer than you might think, because air resistance has an increasingly big effect as the skydiver approaches terminal velocity.

The Meaning of *Zero-Gee* and *Newton*

When someone in a jet airplane, a space capsule, or some other vehicle experiences a state of weightlessness, we sometimes describe the experience as "zero-gee," which can also be written $0g$. Standing still on Earth, that same person experiences "one-gee" ($1g$). Sitting in a steeply banking jet airplane or in a rocket ship accelerating into space, that person may experience three or four gees. Although the word *gee* (and the abbreviation g) is derived from the word *gravity*, none of these situations are measurements of gravity. Instead, they are measurements of acceleration forces, measured relative to gravitational acceleration at sea level (which is given a value of 1). This allows us to compare the net acceleration forces that people experience in these situations to what they would normally experience on Earth. In other words, the g actually stands for "gravitational-acceleration equivalent," not "gravity." To experience "zero-gee" is to be weightless, but the term does *not* necessarily indicate a lack of gravity.

In physics, several different units can be used for measuring force. An especially popular unit is the newton (named for you-know-who), which is the amount of force required to accelerate 1 kilogram of mass by 1 meter per second per second. It is tempting to say that the force of gravity at sea level is equivalent to 9.8 newtons, because gravity will accelerate a 1-kilogram mass by 9.8 meters per second per second. (Note that 9.8 meters is the same as 32 feet, and we previously said that gravitational acceleration is 32 feet per second squared.) The tricky part about this comparison is that while gravitational acceleration at sea level has a constant value, the force exerted by gravity is proportional to the mass of the object, which is why a more massive object is heavier. If you try to lift a 10-kilogram weight and a 1-kilogram weight by a distance of 1 meter (thereby working against gravity), the 10-kilogram object clearly requires more force. Yet if the two objects are dropped, gravity will accelerate them toward the earth at identical speeds, indicating that gravity is applying more force to the heavier object.

The upshot is that a "gee" as a unit of acceleration is able to ignore the mass of an object, but if we think of gravity as a force, we have to take the mass into account. The force of gravity at sea level is actually 9.8 newtons *per kilogram*. Therefore a 10-kilogram weight experiences 98 newtons of gravitational force, and a 100-kilogram weight experiences 980 newtons.

It may be worth noting (or perhaps worth nothing) that a fig newton is a completely unrelated concept. However, whenever I am about to depart on a long journey, I find it helpful to pack a 1-kilogram mass of fig newtons.

Orbits Are Not Always Circular

Up to now, this discussion has treated the orbits of planets, moons, and satellites as if they were circles, maintaining a constant distance from a large gravitational body. Many of our artificial satellites do follow nearly circular orbits, and many planets and moons follow orbits that are roughly circular. But there is no requirement that an orbit be circular. Comets, for example, are famous for having orbits that are extremely elongated. A typical comet spends most of its time far beyond Neptune's orbit but periodically dips into the inner solar system.

However, the path of a comet is not some arbitrary elongated shape; it is always an ellipse. Furthermore, the orbital paths of all objects in the solar system are ellipses. A circle is just a special example of an ellipse, in the same way that a square is a special example of a rectangle.

When a planet, moon, or satellite orbits in a perfect circle, it obviously maintains a constant distance from the larger object that it circles. It also maintains a constant speed in its orbit. But these assumptions break down when the orbit is a highly elongated ellipse, such as the path of a comet. Not only does the distance between a comet and the sun vary considerably during a single orbit, but the speed of the comet also changes dramatically—fastest when the comet is closest to the sun and slowest when the comet is farthest from the sun.

When an orbit is nearly a circle, but not quite, the orbital speed and distance change only slightly during the orbit. For example, most of us consider the earth to travel in a circular orbit, at a distance of 93 million miles from the sun. But in fact, the orbit is not a perfect circle. The earth is closest to the sun in early January (during winter in the Northern Hemisphere), and farthest from the sun in early July (during the northern summer). The distance between the earth and the sun ranges from 91.4 million miles (in early January) to 94.5 million miles (in early July). As a result, the earth travels faster in its orbit in January than it does in July.

We could easily put a satellite into an orbit that follows a clearly elliptical path instead of a circular path. However, we usually find it more useful to put our satellites into circular orbits. For example, the orbit of the space station is very close to a circle.

Where Does Space Begin?

Because we know that gravity reaches deep into space, we can't use a lack of gravity to define where space begins. Contrary to what people sometimes say, going into orbit does not allow you to escape Earth's gravity. So how do we decide where space begins? Can we define outer space as beginning where Earth's atmosphere ends?

As a loose concept this idea works fairly well, but as a precise definition it fails miserably. The problem is that as you fly up into space, Earth's atmosphere gets thinner and thinner, but there is no definite point where we can say that the atmosphere ends. Even the space station, at an altitude of 240 miles above the earth, frequently bumps into stray molecules from the atmosphere. This means there is a tiny amount of atmospheric drag even up that high.

The fact is, no physical phenomenon provides a clear guide to where outer space begins. The best we can do is to define some arbitrary altitude as the official demarcation. And this is exactly what people have done. By international convention, we define outer space as beginning 100 kilometers (62 miles) above sea level. But in reality, when you cross that boundary, you won't notice any sudden changes — and certainly no sign that says Now Entering Space.

I think about these things as I finish lining up my coffee mugs. For instance, I see the Star Trek mug that says "Space: the final frontier." Perhaps it ought to say "100 km altitude: the final frontier." But no — this just doesn't have the same punch. Then there is the mug that shows Newton contemplating an apple in his hand. Instead of holding that apple, he should be tossing it; otherwise, how is he going to obtain any insights into gravity? But what I would really like to find is a coffee mug depicting Einstein in an elevator. I haven't found one yet, but I'll keep looking. If you happen to see one somewhere, be sure to let me know!

3

Survival of the Fittest

I t was laundry day again, and I was hanging the newly washed shirts in my closet. I made sure that every shirt was facing in the same direction—to the left—and that the shirts were organized by color into a continuum, much like a rainbow. When the hanging and arranging were complete, I stepped back to admire the aesthetic appeal of the result. But as I looked at the shirts, I realized there were three I hadn't worn in a long time. I don't wear them anymore because they've gotten a bit too tight. I've been reluctant to get rid of them in hopes that I will slim back down to my former girth. "I've got to do more exercise," I told myself. "If I were more fit, those shirts would fit!"

This unleashed a flood of thoughts. Perhaps I should bite the bullet and just get rid of the shirts that no longer fit. I could call that process "survival of the best fitting." On the other hand, if I were indeed to get more exercise, not only might the old shirts fit again, but also I would probably live longer. I could think of that as "survival due to fitness." But when I hear the popular phrase "survival of the fittest," I have to wonder what the phrase really means. These words are supposed to summarize the concept of biological evolution (also called Darwinian evolution), but do they really accomplish that goal?

What Does *Fittest* Mean?

In our popular terminology, we think of being fit as having a trim body toned by exercise and proper diet. If you were tasked with picking the fittest person in a group of people, you would undoubtedly look for someone who showed the physical signs of maintaining healthy habits. It's easy to imagine that such a person might live longer than others in the group. But taking care of one's health is not the meaning behind the phrase "survival of the fittest."

Of course, we also associate being fit with being strong. So when we hear the phrase "survival of the fittest," we might think this simply means survival of the strongest. By that measure, we can assume that lions will continue to survive, while rabbits and mice are doomed to extinction. However, the world contains far more rabbits and mice than lions. The word *fittest* in this context is not actually a reference to strength, although in some circumstances strength can be a factor that contributes to survival.

Instead, we should examine a different meaning of the word *fit*. We sometimes use the word to mean "appropriate," as in "a meal fit for a king" or "choosing a tool that fits the task." If we think of fitness in this way, it helps us to understand the phrase "survival of the fittest." At any given time and place, the organisms most likely to survive are those that exhibit traits appropriate for dealing with the current circumstances. For some creatures, strength is indeed an essential trait. But for many other creatures, an ability to hide is more valuable. In fact, there are potentially millions of distinct traits that might increase the odds of survival. Therefore, "survival of the fittest" simply means that the organisms with the best combinations of traits for dealing with their current environment are the most likely to survive.

One example of "survival of the fittest" is this: when two species are in competition for the same resources, the species that is more fit will usually win out. While competition between species is certainly an important factor in evolution, an even greater factor is the survival of the fittest *within* any single species. Thus, a key idea behind the phrase "survival of the fittest" is that the individual organisms that make up a species are not all identical. Instead, variation exists among the individuals, and each dissimilar trait might increase—or decrease—the odds of survival.

However, the real issue is not survival but reproduction—generating offspring. The organisms that are better at surviving are far more likely to have offspring, allowing them to pass along their traits to those offspring. The offspring that inherit these beneficial traits are also more likely to survive. Over time, these traits will spread through the population, becoming more and more common, displacing any traits that reduce fitness.

How do organisms pass along their traits to their offspring? The obvious answer is that most of these traits are passed along through the genes—although 150 years ago, at the time of Charles Darwin, this was not an obvious answer at all. But now we know that our genes are encoded in our DNA, and that these genes are inherited through the generations. That said, we should keep in mind that not every trait is heritable. For example, some beneficial traits fall into the category of learned behaviors. These traits are passed along through social contact, especially from parent to offspring. In any species that exhibits learned behavior, cultural evolution as well as genetic evolution can take place. But for the purpose of this discussion, we will focus exclusively on genes.

So this gives us still another way to define *fitness*—as having a beneficial combination of genes that increases the odds of survival.

Gene Variants and Populations

In the popular depiction of how genes operate, we often assume a one-to-one correspondence between genes and traits. In this mental model, each gene controls one trait, and each trait is controlled by one gene. Thus, we might talk about a gene that produces blue eyes, or a gene that makes you tall or short. But in fact, most traits are influenced by several different genes and often by a great number of genes. Furthermore, it is quite common for a single gene to influence more than one trait.

This certainly complicates matters. If survival depends on a great number of traits, and if traits are usually the result of multiple genes, we should not assume that a tiny handful of "good genes" is the key to survival. Instead, a huge number of beneficial genes must all be present. At the same time, a single "bad gene" can sometimes be catastrophic. A bad gene is usually just a slightly incorrect version of a very important good gene. In many cases, the bad gene doesn't directly cause any harm, but because it's faulty, it fails to perform the essential role the good version usually handles. A common result is that the body fails to produce a particular enzyme that's essential for health or growth. This is what we call a genetic disease or disorder.

Most organisms large enough for us to see with our naked eyes (such as humans, grasshoppers, and dandelions) have two copies of every gene. In most cases, one copy is inherited from the mother and the other from the father. These two copies are not necessarily identical. Instead, they can be two variants of the same gene. In the case of most genetic diseases, the disease does not appear if you have one good copy of the gene. The disease shows up only if you inherit two bad copies of the same gene—one from each parent. For example, sickle cell anemia is a genetic disease that afflicts people who have inherited two copies of a specific bad gene.

But if this gene is bad—that is, faulty or harmful—why does it still exist? Due to survival of the fittest, shouldn't the sickle cell gene variant have gradually disappeared, having been outcompeted and replaced by a much better version of the gene? The surprising answer is that people with a single copy of this gene are better able to survive malaria than people who lack the sickle cell variant. Therefore, in places where malaria is common, having a single copy of this gene *increases* fitness, while having two copies of the gene *decreases* fitness. For a population of people living in a malaria-prone area, the presence of the gene in the population increases its overall fitness—that is, its ability to survive.

The example of the sickle cell gene points to the important role of local populations within a species. If the sickle cell gene had first appeared in a part of the world where malaria did not exist, the gene would have quickly disappeared. But the gene survived in a place where it offered an advantage to the population as a whole (even while harming anyone with two copies of the gene).

The word *population*, in the context of biology, refers to a set of individuals that are members of a single species and occupy the same general location; as a consequence, these individuals exhibit a high degree of gene mixing. Physical barriers—such as mountain ranges and bodies of water—often separate the various populations of a species, slowing or even eliminating the flow of genes between two populations. Some of the genes that increase fitness in one population can differ from the genes that improve fitness in another population. This can cause the genetic makeup of two populations to diverge over time. You could say that evolution acts directly on populations and less directly on species. As a general rule of thumb, an important evolutionary change that spreads through an entire species first takes root within a specific population of that species.

Another interesting example in humans is the set of genes that affects skin color. In places that receive a great deal of sunlight, dark skin increases fitness by providing protection from the intense rays of the sun. Light-skinned people who move to the tropics have a high risk of developing sunburns and skin cancer because they lack this protection. But in places that receive much less sunlight, such as the northern parts of Europe and Asia, light skin increases fitness because it helps the skin to produce an adequate amount of vitamin D despite the fainter light.

The upshot is that genetic fitness depends not only on specific genes but also on specific local conditions. A gene that offers an advantage in one place might not be advantageous in a different place. Consider, for example, a species of grass. A particular gene might offer an advantage in habitats that suffer frequent drought, while another gene might offer an advantage in places with soggy soils. If this species of grass expands beyond its original home, it will likely encounter conditions that differ from those back home.

These new conditions can affect which genes are advantageous and which are not, causing a genetic shift in the population colonizing the new location.

Even when a population remains in a single location for a very long period of time, it will typically face an ever-changing set of conditions. The climate in any given location tends to change over time, if you consider a very long time frame. Over a shorter time frame, various catastrophic disruptions (such as floods and fires) can affect a particular location. But the biggest change affecting an ecosystem is often the changing mix of local species. New species move in, and existing species disappear. This results in new predators, new prey, and new competitors. Furthermore, all of these predators, prey, and competitors evolve over time, making use of new genes that increase their fitness for surviving in the current conditions.

The net result is an arms race in which survival depends on adapting to the changes in neighboring species. If your principal source of food improves its defenses against being eaten, your species needs to adapt to this change. If a new species moves into your area, competing for your food supply or nesting locations, you have to adapt. If your worst predator suddenly becomes more efficient at catching you, you have to adapt. All of these changes in local conditions are likely to affect the balance of genes in your own population.

And Now a Few Words About Genes and Mutations

Let's pause a moment to clarify a few thoughts about genes. Your genes are encoded in your DNA, most of which is located in the nucleus of your cells. (A very tiny bit of additional DNA is located in the mitochondria, the organelles within each cell that produce energy.) Generally speaking, every cell in your body contains an identical copy of this DNA. In each cell nucleus, the DNA is divided into chromosomes—in the case of humans, twenty-three pairs of chromosomes. Each chromosome contains, on average, about a thousand genes.

A chromosome is a continuous strand of DNA, consisting of a long sequence of base pairs. Four base pairs—which we abbreviate as G, C, T, and A—are possible. When we do genetic testing, we learn the sequence of base pairs in specially selected parts of the DNA. (These parts are often called genetic markers.) If we sequence the base pairs of an entire gene, we can learn how a particular gene is spelled, using the four corresponding letters. If the testing process is more thorough, we might learn the sequence of base pairs of an entire chromosome. If we learn the sequence of base pairs in all twenty-three pairs of chromosomes, we have sequenced an entire human genome—about three billion base pairs in all.

In one sense, all humans have the same genes; that's what makes us human. Each gene is located on a specific chromosome, at a specific location on that chromosome. Scientists have given individual names to many of these genes. For example, the gene HBB is located on chromosome 11, and its role is to create a protein called beta globin, a crucial component of hemoglobin, which is essential for moving oxygen around the body. Every one of us has two copies of the HBB gene because we all have two copies of chromosome 11. So if humans are characterized by having a specific set of genes arranged in a specific order on twenty-three pairs of chromosomes, what makes us genetically different from one another?

The answer is that most human genes are available in multiple flavors. Well, *flavors* isn't quite the right word—it's better to call them variants or alleles—but you get the idea. When we say that two people have different genes, what we really mean is that they have different variants of certain genes. Many of these variants differ from each other by only one or two base pairs, even though a typical gene consists of thousands of base pairs. For example, the sickle cell variant of the HBB gene differs from a normal HBB gene by only one base pair out of more than four hundred in this relatively small gene.

Such a tiny difference is typical when a new gene variant arises through mutation. This tiny difference might have no effect whatsoever, or it might have a minor effect (either harmful or beneficial) or even a dramatic effect. More often than not, a seriously harmful mutation will cause a zygote to be inviable, so it never develops into a mature fetus. Thus, most of the worst mutations are filtered out promptly.

For many people, the word *mutation* suggests something awful, a dreadful phenomenon that is both harmful and unnatural. In countless bad movies, the world is terrorized by a monster that arose from a mutation, typically attributed to exposure to radiation or toxic chemicals. But many mutations are neither harmful nor unnatural. Spontaneous changes frequently occur in genes, primarily due to copying errors. Almost everyone possesses bits of DNA that differ from those of either parent due to mutations. Our popular culture is slowly catching up with this idea. Recent films are more likely to cast a "mutant" in a heroic role rather than as a mindless monster. This mutant invariably possesses an amazing superpower. However, in real life, our few personal mutations typically have no discernible impact at all.

And yet this accumulation of mutations has a powerful long-term effect on survival of the fittest. Even if the pool of available genes in a population never changed, the relative frequency of the various genes *would* change, partly due to random factors but mostly in response to changing conditions. However, the pool of available genes *does* change, primarily because of mutations and by the introduction of genes from neighboring populations. (At the same time, certain other gene variants gradually go extinct.) While most of the new mutations are either neutral or harmful, some are genuinely useful, increasing the fitness of the individuals that possess the new gene. These beneficial new gene variants are likely to spread through the entire local population, and in some cases, throughout the entire species.

While most genetic mutations simply produce new gene variants, other mutations can have larger effects on the genome. One such outcome is for a stretch of DNA to be omitted, which is usually harmful. Another possible outcome is for a stretch of DNA to be duplicated within the chromosome, which is often harmless (and can be beneficial in the long term). Such a mutation is often passed down through the generations. A third possible outcome is that an entire chromosome is duplicated. In humans, this is usually harmful, resulting in genetic disorders such as Down syndrome. But in the plant kingdom, instances in which the entire set of chromosomes is duplicated—a condition called polyploidy—are quite common and usually beneficial. Examples include several of our commercial crop plants, such as oats, peanuts, sugar cane, bananas, potatoes, tobacco, cotton, strawberries, and rutabagas. Instead of having two complete sets of chromosomes, a polyploid plant has three, four, six, or eight complete sets of chromosomes in the nucleus of each cell.

Over time, these various kinds of large-scale genetic changes can cause related species to diverge as to the number of chromosomes or to differ in the sequence of genes within a chromosome. Approximately seven million years ago, a species of ape gave rise to both humans and chimpanzees. Today the genes and chromosomes of humans and chimpanzees are still quite similar. But one important difference is that chimpanzees, like most other great apes, have twenty-four pairs of chromosomes, not twenty-three as in humans. How did that happen? In the human line, two of the formerly independent chromosomes fused end to end, becoming a single chromosome. Even today, human chromosome 2 is largely identical to a combination of chimpanzee chromosomes 2A and 2B.

Putting It All Together

We have now covered eight key ideas that explain the concept of biological evolution:

1. A species is never completely uniform. Instead, it consists of individuals that possess varying traits.
2. Certain traits increase the odds of survival. Individuals that possess these traits are likely to pass them along to their offspring. As a result, such traits spread through the local population.

3. Most of these beneficial traits are influenced by genes. Genes that increase the odds of survival spread through the local population.
4. Mutations constantly introduce new gene variants into a population. At the same time, certain gene variants—some new, some old—are eliminated from the population. Thus the mix of available genes changes over time.
5. Local conditions change over time and vary from place to place. This variation in conditions has a huge effect in determining which genes are currently advantageous to a population of organisms.
6. While climate is a crucial factor affecting local conditions, it is not the sole factor. For example, the environment for any species is also highly influenced by the other species that share the same ecosystem.
7. Evolution is driven by the survival of the fittest in response to current local conditions, based on traits influenced by genes that are currently present in the population.
8. In some cases, beneficial new genes eventually become ubiquitous in the species. In other cases, separate populations diverge to the point of becoming distinct subspecies and eventually distinct species.

"The Ascent of Man" and Its Fallacies

When people think of evolution, they often focus primarily on ideas about human evolution. In the popular imagination, the essence of evolution is summarized by the image of a parade of creatures marching left to right, with a modern man at the head of the parade and something resembling a gorilla or chimpanzee bringing up the rear.

This parade, often called "the ascent of man," is striking and memorable—and deeply embedded in our culture. Many artists and

A familiar stereotype of human evolution

illustrators have created their own versions of the image. Because of its high degree of familiarity, this stereotype has spawned a great number of humorous memes. Most people—even people who don't believe in evolution—assume that this picture accurately communicates the basic concept of evolution. Unfortunately, the image is highly misleading because it promotes ideas that conflict with our modern understanding of evolution. This picture has two principal problems: the fallacy of the linear path and the fallacy of the higher state.

Contrary to the story conveyed by this image, the evolution of humans was not linear. Instead, the pathway from ancestral apes to our current form is a tree with many branches. For most of the long history of human evolution, several species of hominins existed simultaneously. (The term *hominin* refers to members of the genus *Homo*—such as humans and Neanderthals—along with slightly more distant relatives such as members of the genus *Australopithecus*.) This sharing of the world by a variety of hominins continued for several million years. It was only about thirty thousand years ago that hominins dwindled to a single species—*Homo sapiens*. Consequently, when we find ancient hominin fossils, it can be difficult to determine if these are our direct ancestors or if instead they represent dead-end evolutionary branches—closely related to us but not directly ancestral.

Although our understanding of the hominin family tree is still incomplete, paleontologists continue to find new fossils that paint an increasingly clear picture. Based on our current knowledge, our family tree (including the extinct hominins and the most closely related apes) probably looks something like the diagram on the opposite page.

It's not surprising that the general public still tends to picture evolution as a linear path. Only a few decades ago, our textbooks and museums often depicted human evolution as a simple unbranched line. In addition to the evolution of humans, another popular example was the evolution of horses, stretching from *Eohippus* to the modern horse. This, too, was typically depicted in a completely linear fashion—and unfortunately, this example was equally misleading. Horses, like humans, have many branches in their family tree.

But even while textbooks and museums displayed these misleading representations, the famous example of Darwin's finches communicated a very different story—the story of adaptive radiation. Fourteen species of finchlike birds, all descended in just a few million years from one common ancestral species, live on the Galapagos Islands. These fourteen species differ in several physical traits, the most obvious being the shape of the beaks and the size of the bodies. These differences have allowed the various species to specialize in different sources of food. Each new species was able to gain an advantage by utilizing a resource that had been underutilized before.

Evolution quite frequently occurs in episodes of adaptive radiation, where an existing species gives rise to several new species that coexist, at least for a while. (This is especially common when a species finds itself in a time and place with a paucity of other species and an assortment of unoccupied ecological niches available for exploitation.) Each of the new species may in turn generate several new species. At the same time, a constant process of extinction means that existing species disappear. This is reflected in our evolutionary tree diagrams, where so many branches and twigs eventually lead to dead ends. These two processes—evolution and extinction—operate simultaneously, resulting in an ever-shifting balance of species in the world, especially when viewed in a time frame consisting of millions of years. The result is certainly not linear.

Because we so often imagine evolution as a linear path, it's easy to believe that evolution represents a journey from a lower state to a

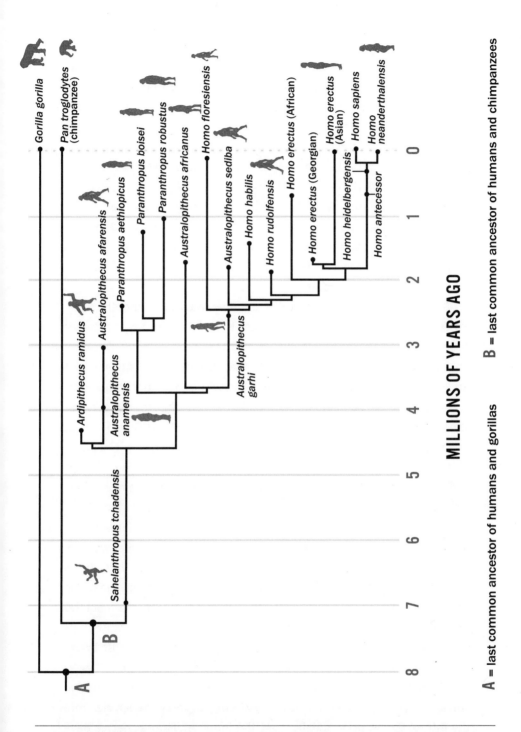

MILLIONS OF YEARS AGO

A = last common ancestor of humans and gorillas   B = last common ancestor of humans and chimpanzees

A more accurate depiction of human evolution. Note that the diagram does not show the many branches within the family trees of gorillas and chimpanzees. (Adapted from a 2018 preprint version of Caroline Parins-Fukuchi, Elliot Greiner, Laura M. MacLatchy and Daniel C. Fisher, "Phylogeny, ancestors, and anagenesis in the hominin fossil record," which later appeared in *Paleobiology* 45, May 2019, 378–393.)

higher state. In this popular view, the end result of evolution is a creature that is superior to its ancestors. The "ascent of man" illustration certainly reinforces this concept—and the word *ascent* communicates this message quite overtly. But in reality, evolution simply produces species that are better adapted to survive current local conditions. Nothing in this process requires a higher state to emerge. All species evolve, but it would be hard to argue that the grasshoppers and salamanders in today's world represent a higher state than the grasshoppers and salamanders of five million years ago. Bacteria tend to evolve quite rapidly, and they have been doing so for several billion years, but no one suggests that today's bacteria represent a dramatically higher state as a result.

Of course, if we think in terms of adaptive radiation rather than linear evolution, we're not likely to view evolution as a journey to a higher state. Few people would argue that all fourteen species of Darwin's finches have achieved a higher state than the ancestral species that first colonized the Galapagos Islands. And while the concept of evolving to a higher state is deeply embedded in our culture, the concept itself is poorly defined. What exactly is a higher state? Does it mean increased intelligence? Increased size? Increased strength? Examples indeed exist of species in which increased intelligence, size, or strength has proven to be an adaptive advantage. But far more examples can be found in which none of those attributes were involved.

A common trope in movies and TV shows is the scientist who invents a way to speed up evolution. As the drama unfolds, we may see a brave or unwitting volunteer step into a chamber that sports dials and flashing lights. The scientist makes a few adjustments to the control settings and then starts up the machine. A few minutes later the volunteer steps out, having evolved by millions of years. Perhaps the volunteer now has a huge head to accommodate a gigantic brain. And of course, the volunteer has now acquired several superpowers.

This vision of scientific progress is, of course, riddled with errors. No single individual has all the necessary genes to carry the species into the future. Evolution works because the individuals in a species have sets of genes slightly different from one another, providing a huge pool of potentially useful characteristics. But the most critical flaw in this scenario is the idea that the evolution of a species is predetermined by an existing, linear track to a future higher state—and that you simply need to hop onto that track and turn up the speed. This ignores the crucial influence of unpredictable future mutations on a population's gene pool. It also ignores the unpredictable future conditions that the population will face (including the evolution of other local species). Thus, we cannot predict the future evolution of any particular species, nor can we predict how that species might radiate into multiple species. No preordained linear track to a future higher state exists.

Consider another example: our use of the word *devolve*, which has become quite trendy in recent years. Every few days I encounter the word in another news report or opinion piece. The authors use the word to mean "decline," "degrade," "descend," "degenerate," "decay," or "regress." For example: "The congressional committee hearings have devolved into political theater." Or: "Portions of the country may soon devolve into chaos." This use of the word is based on the idea that *devolve* is the exact opposite of *evolve*—and that evolution is a linear journey to a superior state. To devolve, therefore, is to descend into an inferior state. This use of *devolve* betrays a deeply embedded popular misunderstanding of biological evolution.

As a final nail in the coffin of the "higher state" concept, consider the example of whales. We tend to think of whales and dolphins as advanced creatures because they are large and intelligent. But the evolution of whales has involved a great number of traits, not just size and brainpower. Whales are mammals, so they nurse their young with milk and they breathe air using lungs. Mammals descended long ago from fishes, which lived in the ocean and propelled themselves with fins. As certain fish found opportunities on solid ground, the fins gradually evolved into legs, giving rise to all sorts of land-based vertebrates, including mammals. But later, as certain mammals found new opportunities back in the ocean, the legs became flippers, quite similar in appearance and function to fins. Does this mean that whales devolved to an earlier state? Is it a higher state to live on land or to live in the ocean? Is it a higher state to have legs or to have flippers? The reality is that land and water both provide opportunities as a place to live, but that legs are quite useful on land, while flippers are quite useful in water.

"Descended from Monkeys"

Another phrase commonly used by the public when discussing evolution is "descended from monkeys." Of course this is a human-centric perspective, reflecting the idea that if you trace human ancestry back far enough, you will reach a species that is decidedly not human. Unfortunately, this bit of shorthand tends to promote two misconceptions: that humans are descended from existing species of monkey or ape, and that human evolution began with monkeys or apes. The following paragraphs set the record straight.

Humans are not descended from any existing species of ape or monkey. Instead, humans share common ancestors with apes and monkeys. (Technically, humans *are* apes.) These common ancestors are no longer around. All apes and monkeys have continued to evolve, causing them to differ from those common ancestors, just as humans have continued to evolve. However, the last common ancestor shared by humans and chimpanzees is different from the last common ancestor for humans and gorillas because we are more closely related to chimpanzees than to gorillas. All apes and monkeys share a complicated family tree with many branches. Each branch point represents a common ancestor, but it also represents a point of divergence between two species. (In some cases, two closely related branches continued to interbreed for a while, further complicating the picture.) The last common ancestor shared by humans and chimpanzees lived about seven million years ago, while the last common ancestor for humans and gorillas lived between eight and ten million years ago. Our last common ancestor with true monkeys lived about twenty-five million years ago.

Human evolution goes back much farther than monkeys. Monkeys and apes share a common ancestor with all other primates (such as

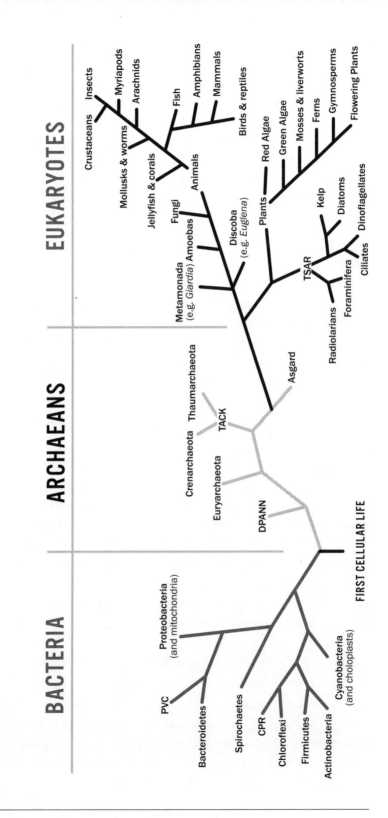

The long history of evolution (highly simplified)

lemurs). Primates share a common ancestor with all other mammals. Mammals share a common ancestor with all other vertebrates. As you trace our vertebrate ancestors farther and farther back, you eventually find that all vertebrates are descended from fish, although these ancestors are not the same species of fish that live today. But if you go back even farther, you find that fish had ancestors that were definitely not fish. A core concept of evolution is that our ancestry goes all the way back to simple one-celled organisms. The same is true for all other living things on Earth—animals, plants, fungi, bacteria, and everything else. All life on Earth is descended from simple one-celled organisms that lived well over a billion years ago.

This history is reflected in our DNA. All living creatures use the same molecule—DNA—to encode their genes. All living creatures use the same DNA encoding system, consisting of sixty-four three-letter codes. The DNA in all living creatures encodes for the same set of twenty amino acids (with a few rare exceptions). These amino acids serve as the building blocks for proteins, producing the thousands of different enzymes that direct growth and metabolism, thereby controlling the traits of each organism. To a surprisingly large degree, humans share many genes with many other living creatures. Even simple creatures such as sponges and worms have crucial genes that are amazingly similar to certain human genes. It is because of this striking similarity that we can now study DNA to piece together the family trees of all living creatures. This approach usually produces a more detailed history than using fossils alone. The process of creating these family trees is time consuming and not yet complete, but current progress is amazingly rapid. As a result, we gain a deeper understanding of evolutionary history with each passing year.

After my long daydream regarding the details of evolution, my thoughts returned to the arrangement of clothes hanging in my closet. I soon began to reflect on how my wardrobe has evolved over the years. In any given year, my collection of clothing has been similar to the collection I had a year earlier. Each year a few new items appear and a few old items disappear, but the changes are seldom massive or abrupt. And yet, over time my wardrobe has evolved considerably, gradually adapting to changes in my taste, girth, and typical activities. Of course, this evolution has always been constrained by the styles of clothing currently available for purchase—much like biological evolution is constrained by the set of genes that are currently available.

I closed my closet door, content with the current arrangement of items, which seemed to me both logical and aesthetically pleasing. I could finally turn my attention to rearranging the clothes in the drawers of my dresser.

4

The Five Senses

I often find myself driven by a sense of aesthetics, as when I organize the shirts in my closet by color or smooth the surface of the yogurt as I eat it. On the other hand, I sometimes perceive a conflict between aesthetic and practical considerations. When I encounter such a conflict, I usually take the side of the practical.

Consider the example of a roll of paper with tear-off segments, such as paper towels hanging in the kitchen or toilet paper hanging in the bathroom. When the roll is positioned so that the loose end hangs behind the roll, next to the wall, the result definitely has a pleasant appearance. But when the roll is positioned so that the loose end hangs in front of the roll, it's easier to grab the paper and to find the end if it's not hanging down. Thus, I think the only correct way to hang toilet paper is to position the loose end in front. In fact, this seems quite obvious to me. The upshot is that it drives me crazy whenever I see a roll of toilet paper hanging the "wrong" way. I can clearly see that the roll needs to be turned around. Whenever I encounter this situation, I am seriously tempted to correct the problem. All too often, I find myself giving in to that temptation.

A recent encounter with a misaligned roll of toilet paper led me to contemplate these conflicting senses. On the one hand, there is a sense of beauty and aesthetics. On the other hand, there is a sense of practicality, reason, and order. We sometimes summarize the latter by saying, "It just makes sense!" Someone who practices this latter approach is said to be sensible. But someone who prefers the former approach—that of beauty, aesthetics, and emotions—is often said to be sensitive. So it seems that regardless of how we approach these matters, we want to attribute our thinking to senses.

However, this use of the word *sense* is quite broad. It can apply to anything going on in your head, especially if it relates to an emotion or attitude. We can talk of a sense of humor, a sense of wonder, a sense of entitlement, a sense of loss, or a sense of honor. In fact, we have dozens of common phrases in English that begin with "a sense of." These phrases tend to describe how you might respond to situations around you. But what if we focus on how you *detect* things rather than how you respond to them? This represents a very different sense of the word *sense*. When we discuss the senses in a scientific sense, it is this much narrower meaning—the detection of things around us—that is the subject of our attention.

No Consensus on a Census of the Senses

We traditionally assume that humans have exactly five senses: sight, hearing, taste, smell, and touch. Each of these five senses makes us aware of things outside of ourselves, things in the world around us. This way of categorizing the senses is ancient, dating back more than two thousand years. On the assumption that this model is factually correct, we teach our children about the five senses from a very early age. This model is so ingrained in our culture that any additional method of perception, whether real or imagined, is often called a sixth sense.

However, our traditional model of five senses has some serious weaknesses. By most objective measures, humans actually possess more than five senses. So it seems rather odd that we continue to teach children that we have just five. That's not to say that this old model is completely worthless. Because the model is so simple, it's easily learned, even by very young children. Thus, it can serve as a helpful introduction to the concept of human senses. But for older children and adults, the model seriously constrains our thinking about the senses.

A principal characteristic of the five senses model—and one reason it's so appealing—is that each of the five senses is paired with a highly visible part of the body: eyes, ears, mouth, nose, and skin (often emphasizing the hands). In fact, this way of thinking is actually a model of our five most obvious sense organs rather than a proper model of the senses. And this helps to explain why we feel so tempted to teach the model to toddlers in conjunction with their learning to identify and name the major parts of the head and body.

Unfortunately, there is no universal agreement as to how many senses we humans actually possess. While all of our senses relate to detecting things, this vague definition leaves a lot of wiggle room. Furthermore, the issue of lumping versus splitting affects our final count—that is, when we identify two closely related senses (example: detecting hot and detecting cold), should we consider them two different senses or two aspects of the same sense? If we broaden our consideration beyond humans, our count might include senses some animals possess but humans do not, such as the ability to detect magnetic fields. But the biggest factor in the differing counts is that you have senses for detecting situations inside your own body—such as feeling pain or knowing when your bladder is full—and these additional senses are not directly acknowledged by the ancient model of five senses. (Even among scientists, these internal senses are often lumped into the sense of touch.) For all of these reasons, experts disagree as to how many senses you actually have. Because no consensus exists as to how best to replace the five senses model—or even whether the model should be replaced at all—the old model is able to retain its popularity with the general public.

As mentioned previously, a key characteristic of the five senses model is that all of the senses are related to detecting phenomena that originate outside of your body. In other words, the five traditional sense organs are all tools for investigating the world around you. You see, hear, smell, taste, and touch the things that surround you. If you limit your census of the senses to those that detect external phenomena (ignoring phenomena that originate inside your own body), your count will never get very long—although your final count will probably be greater than five.

What External Phenomena Can We Detect?

All of the external things you are capable of detecting are physical phenomena—in other words, matter, energy, and forces. Your sense organs are capable of detecting various types of matter and energy that come into contact with your body, along with forces that act upon your body (such as gravity). These organs rely on your nervous system to relay the resulting messages to your brain. To that end, each of your sense organs has specialized nerve endings connected to specialized cells called receptors that can detect these various phenomena. Therefore, to count how many senses we actually have, we should begin by itemizing the types of detectable physical phenomena that originate outside the body. (Later, we can look at detectable phenomena that originate inside your body.)

The detectable external phenomena are as follows:

1. light (electromagnetic radiation)
2. sound waves (vibrations)
3. odors and flavors (chemical molecules)
4. direct contact (touch or pressure from contact with matter)
5. heat and cold (temperature)
6. gravity and acceleration
7. magnetic fields
8. electric fields and static charges

Let's look at each of these eight types of external phenomena in more detail.

LIGHT (ELECTROMAGNETIC RADIATION)

Every day, the world around us is bathed in energy from the sun, which we call light. As this light strikes objects and surfaces on the earth, some of this light is absorbed, some is reflected, and some passes through the objects. You use your eyes not to look directly at light coming from the sun, but to see light that has scattered off the objects around you and thus to see those objects. This all works because your eyes can detect light—or more precisely, your eyes can detect a limited range of wavelengths within the spectrum of electromagnetic radiation. We call these wavelengths visible light, which is simply an acknowledgment that we can detect them with our eyes. Because we are so dependent on our sense of vision, we have found ways to illuminate the world around us when the sun is not shining, originally by using fire to produce light and then later by using electric bulbs of various types.

Your ability to see is far more sophisticated than simply detecting the presence of light. The lens in each eye focuses images on the retina, which allows your brain to deduce the shapes of objects that reflect or emit light, as well as the exact direction from you to the object. The retinas of your eyes include four kinds of photoreceptors: rods and three kinds of cones. Each type of cone responds to a different range of wavelengths, allowing your brain to perceive color. The rods are critical for low-light vision. The fact that you have two eyes with overlapping fields of vision provides you with depth perception, enhancing your ability to judge distances. Intense visual processing in the brain allows you to identify faces and edges in your field of vision. Your brain constantly compares the visual input from one moment to the next, which is how it detects motion.

Several kinds of animals, including birds and bees, have the ability to see frequencies of light in the ultraviolet range, which humans cannot see. (We don't consider this to be visible light, even though birds and bees can see it. Apparently, if you aren't human, you don't count!) Certain kinds of snakes can detect infrared light using pit organs in their heads, allowing them to detect the body heat of their prey. The sensors in these pit organs work by detecting subtle temperature changes in the tissue lining the pits rather than directly detecting the photons of infrared light, so pit organs are quite different from eyes.

SOUND WAVES (VIBRATIONS)

Sound waves passing through the air are another detectable external phenomenon, providing you with information about the world around you. Your ears collect and detect sound waves within a certain range of frequencies. Although you cannot hear sounds with frequencies that lie outside that range, humans are very good at distinguishing among the audible frequencies and distinguishing other characteristics of sounds. Because you have two ears, you have a sense of which direction a sound is coming from. These abilities help you to detect what is happening all around you, and they also allow you to communicate with other people through speech.

Sound waves are a type of energy, but unlike light, sound requires a medium (consisting of matter) to transmit the waves. For humans, that medium is usually air, but water can also carry sound waves. The vibrations of sound waves can even travel through solid materials. All of these media provide opportunities for picking up information about the world. In fact, some animals are quite skilled at detecting vibrations in media other than air. Many animals that live in water have an excellent sense of hearing. Other animals can detect vibrations in solid objects. For example, an insect trapped in a spiderweb causes vibrations that not only alert the spider but also tell the spider certain details about what has been caught. Many kinds of animals, including elephants, can detect and interpret vibrations coming through the ground. Some of these examples might not be classified as hearing—lumped instead with the sense of touch—but they all are based on extracting information from vibrations in matter.

Some animals, such as bats, have developed the ability to "see" their surroundings through echolocation. They can determine the precise locations and shapes of nearby objects by detecting sound waves bouncing off them, somewhat analogous to our own ability to assemble a mental image of the world around us by observing reflected light. The bats themselves create the sounds (in the ultrasonic range, beyond the range of human hearing), using the echoes to create mental maps of the surrounding space. We apply the same principle when we use radar or sonar to detect objects and determine their size, speed, and direction.

ODORS AND FLAVORS (CHEMICAL MOLECULES)

Your senses of smell and taste are both based on detecting molecules of various substances that come into contact with your body. In the case of smell, you use olfactory receptors in your nose to detect a wide range of airborne molecules. In other words, you detect substances that are floating in the air. In the case of taste, you have at least five distinct kinds of receptors on your tongue to detect certain molecules in your food.

Your perception of taste is due to input from both of these senses. The taste buds on your tongue detect molecules that you perceive to be sweet, sour, salty, bitter, or savory (also called umami). All of the other flavors that you detect in your food are due to molecules that enter your nose—mostly from the back, because your mouth and nose are connected via the throat. As you chew your food, you release volatile molecules that waft up through this connection into the nose. In contrast to the five distinct types of taste buds, at least four hundred distinct olfactory receptors are present in the nose, each corresponding to a separate gene in your DNA. These four hundred sensations result in millions of possible combinations, allowing you to detect a huge range of distinct odors.

As we all know, many animals besides humans use their noses to smell and their mouths to taste. The surprise is that certain creatures can taste or smell by using other sense organs. Some insects can detect and distinguish airborne molecules with their antennae, meaning they use their antennae to smell. Some insects can distinguish molecules in materials that they touch with their feet, meaning they have a sense of taste in their feet.

We humans are able to detect a limited range of chemicals that come into contact with our skin because these particular chemicals trigger thermal receptors in the skin. Some chemicals (like the capsaicin in chili peppers) make the skin feel hot, while others (like the menthol in mint) make the skin feel cool. However, in our usual models of the senses, we lump our human skin-based chemical sensations into the sense of touch rather than the sense of taste.

DIRECT CONTACT (TOUCH OR PRESSURE)

Another opportunity to gather information about the world around you is to detect matter that comes into contact with your body. You have many distinct kinds of receptors in your skin, several of which specialize in detecting touch or pressure. (To call it *touch* emphasizes that matter has come into contact with the skin, while the word *pressure* refers to the force exerted on the skin by that same matter.) This allows you to determine when your body has encountered an external object. This primarily applies to solid matter and to liquids, but your skin can also detect moving air. Although these receptors are in the skin all over your body, the density of the receptors varies considerably. Some parts of your skin—such as on your hands—have a large quantity of receptors packed into a small area, giving those parts of your skin a much better ability to gather information and to discern shapes, sizes, and textures than other parts of your skin.

The skin on your hands has a second advantage compared to other parts of your skin. The flexibility of your hands allows you to explore surfaces in detail. With your eyes closed, you can easily determine the shape and size of a small object just by touching it with your hands. This is very hard to do with any other part of your skin. Part of the trick is that you don't have to feel the entire surface simultaneously. You can spend several seconds feeling different parts of the surface, and then your brain puts the information together.

The sense of touch can be extended over some distance by the use of a long, slender appendage, such as the whiskers of a cat or the feelers of an insect or crustacean. In the case of a cat's whiskers, the touch receptors are located in the skin surrounding the base of the whisker. But in the case of a feeler (an antenna used for touching), the touch receptors are actually located in the feeler. In many cases, the same antennae contain other kinds of sense receptors, allowing for smell, taste, hearing, or other capabilities.

HEAT AND COLD (TEMPERATURE)

Another category of receptors in your skin detects changes in temperature: the hot and cold receptors. Although this type of receptor provides you with information about the world around you, it does so indirectly because these receptors do not directly sense the outside world; instead, they detect temperature changes in your skin. Skin is heated or cooled by contact with other objects or the air, and also by the exchange of radiant energy (primarily infrared radiation). Therefore when you feel the heat of a fire, it is not by directly detecting the infrared radiation striking your skin but by detecting the resulting change in the temperature of your skin. (The main difference between your temperature sense and the pit organs of a snake—other than the degree of sensitivity—is that the shape of the pit organs allows the snake to more accurately pinpoint the direction from which the energy originates.)

In the five senses model, the sense of hot and cold is traditionally lumped together with the sense of touch because the same sense organ (the skin) is involved in both. (In the scientific literature, any sensation detected by the skin or by organs deeper within the body can be categorized as *somatosensory*, which literally means "sensed by the body.") However, sensing temperature is quite different from sensing contact because it relies on a different phenomenon and different receptors. After all, you don't have to touch the sun to feel its heat! Consequently, models that identify more than five human senses typically include thermoception (detecting temperature) as one of the additional senses.

GRAVITY AND ACCELERATION

Your ability to detect gravity is the principal component of your sense of balance. Gravity is detected by the semicircular canals in your inner ears. This is a very real sense, with a clearly identified sense organ. Yet this sense is not included in our traditional five senses model, in part because the sense organ is not visible on the outside of the body and in part because the five senses model predates our understanding of the role of the semicircular canals.

In some ways this is a rather subtle sense. You don't usually think about it except when it fails you—that is, when you feel dizzy and have a hard time standing up without keeling over. But such a failure reveals the great importance of this sense. Your ability to detect gravity allows you to determine which way is up so that you can maintain your body in an upright position when you stand or walk, even when your eyes are closed. And while we sometimes talk about our sense of balance, most of us fail to recognize it as a real sense. It's also worth noting that certain other senses, including vision and proprioception (which we'll discuss soon), also contribute to your ability to maintain your balance.

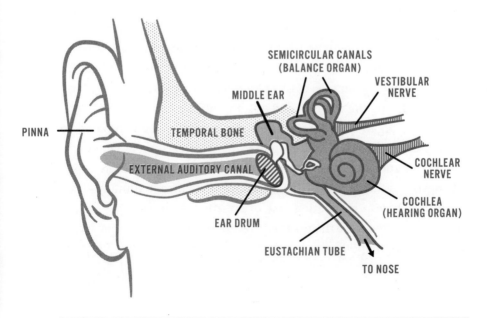

The location of your balance organ

Although gravity is a phenomenon that results from the attraction between any two masses, from a practical standpoint for life on Earth, gravity is the attraction between the earth and all other matter on or near the earth. From the standpoint of physics, it's impossible to distinguish forces that result from gravity from those that result from acceleration. Thus, your semicircular canals cannot tell the difference between gravity and acceleration. If you subject your body to any type of significant acceleration—especially if the direction or amount of acceleration rapidly changes—the information sent from your semicircular canals to your brain can be confusing. If you spin on a rope or ride a boat that's pitching on the waves, the confusing signals can leave you feeling dizzy or queasy.

While mammals rely on their semicircular canals to determine the direction of gravity, many invertebrates use a very different organ called a statocyst. In either case, the purpose is to detect gravity in order to know which way is up so that the body can be properly oriented for safety or locomotion.

MAGNETIC FIELDS

Many kinds of animals are able to detect magnetic fields, even though humans cannot—at least not to any significant degree. The ability to detect the earth's magnetic field tends to give these animals a powerful sense of direction, especially the directions of north and south. The best-known examples of this phenomenon are birds that fly long distances during their spring and autumn migrations.

A sense organ that detects magnetic fields can be compared to a compass. The individual receptors might be extremely small and could theoretically be anywhere in the body, even in the brain itself. The upshot is that while we have excellent evidence that many kinds of animals have a magnetic sense of direction, in most cases we are not sure exactly where the magnetic receptors are located. A few years ago, a leading hypothesis was that the magnetic receptors for birds are located in their beaks. More recently, new evidence has suggested that the receptors might be located in the retinas of their eyes. Perhaps soon we'll have a definitive answer.

ELECTRICAL FIELDS AND STATIC CHARGES

Some aquatic animals have the ability to sense changes in the electrical field in their immediate vicinity with special sense organs embedded in their skin. The best-known examples are sharks and rays, but other creatures with this ability include sturgeons, paddlefish, catfish, elephant fish, and platypuses. This sense can be used to identify prey and other nearby objects, which can be quite useful when it's dark or the water is murky, or when the prey is hiding in mud or silt. Some species, such as elephant fish, even generate their own electrical field, which makes it a lot easier for them to detect nearby objects—sort of like carrying a flashlight. This ability can also be compared to the echolocation of bats, which generate ultrasonic chirps to sense their surroundings.

Water is a much better conductor than air, and therefore most creatures that rely on electroreception are aquatic. However, some land animals—even humans—can detect static charges, often through indirect means. In fact, bees and certain spiders appear to make significant use of their ability to detect static charges. In the case of humans, a nearby static charge causes the hair on your arms to stand up, which you can easily feel. Of course, you can also feel gusts of wind with the hairs on your arms. The receptors surrounding the hairs cannot distinguish between these two phenomena, but your brain, upon receiving the information from many hair follicles over a period of several seconds, can distinguish between the two situations. This ability is typically categorized as part of your sense of touch—like the whiskers of a cat—rather than a separate sense. This contrasts with the specialized sense organs of aquatic animals (such as sharks) that directly detect electrical fields.

The Internal Senses

Let's pause here and take stock of our list so far. We have identified eight detectable external phenomena—nine if you separate airborne chemical molecules (odors) from nonairborne (flavors)—and every one of these phenomena corresponds to a specific sense in various animals. Humans have seven of these nine senses, lacking only the ability to detect magnetic fields and electrical fields. If we define the word *sense* as "the ability to detect external phenomena," our count is finished: there are seven human senses but nine senses across the animal kingdom. However, we could define the word *sense* in several other ways.

For example, your body has additional sense receptors beyond the ones we have catalogued so far. Instead of detecting phenomena external to your body, these additional receptors provide information about the body itself. The most obvious example is your sense of pain, triggered by pain receptors located not just in your skin but also deeper within your body. (Broken bones and other internal injuries can certainly cause pain.) We are also aware of internal phenomena such as being hungry or thirsty, feeling too full from having eaten too much, and having a full bladder. All of these require sensors within the body in order to detect the issue. The receptors for these senses send messages to the brain via the nervous system. Therefore, we could legitimately refer to a sense of hunger, a sense of thirst, or a sense of being full. In fact, scientists have catalogued a long list of such internal senses. If we were to include all of these senses in our list, we could easily reach twenty or thirty distinct human senses.

The sense of pain is triggered when pain receptors (also called nociceptors) react to various stimuli. Some react to excessive pressure or mechanical deformation (the stretching, twisting, or bending of parts of your body). Others react to excessive temperatures—either too hot or too cold. Some detect exposure to certain chemical compounds, while others detect damage to nearby cells. A person's survival can depend on detecting and responding to such stimuli quickly, so our brains have evolved to prioritize pain signals above everything else. The result is that the sense of pain can often be overwhelming. Although pain receptors react to many of the same stimuli as the touch and temperature receptors in your skin, they are a completely separate class of receptors, and they even use a completely different set of nerves to send signals to the brain.

Certain internal senses—separate and distinct from the sense of pain—involve receptors that detect the stretching of body tissue. The stretching of your stomach gives you the sensation of being full, while the stretching of your bladder gives you the sensation of needing to urinate. A different set of receptors detects the tension in your muscles—that is, the degree to which individual muscles are contracted or relaxed. Other receptors, mostly in your skin, trigger an itching sensation.

Still other internal senses rely primarily on chemical cues. If sensors in your stomach detect excessive amounts of certain chemicals, the vomiting reflex is triggered. Sensors elsewhere detect when the amount of water in your body has gotten low, triggering a sense of thirst. A sense of hunger can be triggered by an empty stomach, but a sense of needing food can also be triggered by a drop in your blood sugar level.

Other receptors deep in your body respond to temperature. These receptors have a different purpose from the thermoreceptors in your skin, and they generate a different set of reactions. These internal receptors are involved in thermoregulation, keeping the core of your body (especially your brain) at a consistent, optimal temperature. If these sensors detect that your internal temperature is getting too high, they set off a suite of reactions to cool your blood by increasing your sweating and opening up the capillaries to move blood closer to your skin where it can lose heat. If these sensors detect that your internal temperature is getting too low, this sets off a different set of reactions, such as goosebumps and shivering.

One other internal sense, a subtle but important one, is called proprioception. This is the sense of knowing how the various parts of your body are positioned without relying on sight or touch. A demonstration of this sense is to close your eyes, then reach up and touch your nose. Most people can do this quite easily. Likewise, you can easily reach out to scratch an itching ankle without looking, regardless of whether you're standing up, sitting in a chair, or lying curled up in bed. Proprioception relies on flex sensors located in your muscles, tendons, and joints. These receptors don't actually detect where in space your limbs are located, but the information supplied by these sensors allows your brain to deduce the locations of your body parts. Proprioception also plays a role in your sense of balance, providing information beyond what is provided by the semicircular canals in your inner ear.

Many articles in the popular press have stated that this newly recognized sense means that humans actually have six senses instead of five. This is a rather dubious claim given that several other senses are also worthy of joining the expanded list. In fact, I would argue that the strongest candidate for the sixth slot is the sense of balance, which has its own distinct sense organ (the semicircular canals) and is triggered by a distinct external phenomenon (gravity). The senses of temperature and pain are also strong candidates for joining the list because of their unique characteristics, and proprioception should probably be included too.

A brief aside about terminology: for all of the traditional senses, our vocabulary includes at least one popular term and at least one technical term. For instance, we can refer to *smell* or *olfaction*, but either way we're discussing the same thing. In a college course about the senses, the terminology often skews heavily toward the technical side. But to discuss the senses with small children—or even in casual conversation with adults—we instead employ the popular terms.

In the case of the newly recognized sense of proprioception, we have a formal term but no clear agreement on a popular term. If we intend to teach young children about this sense, it would be helpful to employ a simple, familiar word, as we do with all the other senses. But what would that word be? Some people explain proprioception as knowing the location of one's limbs, but "a sense of location" would be a highly misleading phrase. Proprioception can also be explained as knowing the position of your body in space and the position of each of your limbs. Thus, a good popular term for this sense would be *a sense of position*, although some people have also used the term *a sense of space*.

Perception

This brings us to the concept of perception, the mental processing of the information your brain receives from the various sense receptors. For each human sense, we can make a distinction between what the sense organ actually detects and what the brain perceives. In the case of proprioception, your brain converts information about the flexing of your muscles and joints into a mental map of where your limbs are located and how they are positioned. In the case of vision, your brain takes the raw data from the eyes and converts it into a perception of shapes, colors, sizes, distances, and movement. The brain puts a lot of work into converting sensory information into perceptions, and most of this work occurs automatically without your being aware of it.

In all of our senses, detecting something with the sense organs is only the first step. The information then needs to be relayed to the brain via nerve pathways, many of which are specialized for carrying distinct kinds of information. The brain assembles and interprets the information it receives to produce your perception of the sense. When we use the term *human senses*, we tend to focus on the sense organs and how they detect various phenomena. When we use the term *perception*, we instead focus on how the signals from the sense organs are interpreted by the brain. Either way, your sense organs, nervous system, and brain are all involved. But discussing senses instead of perception can cause us to underappreciate the huge role of the brain in these processes.

In humans, a large part of the brain is devoted to processing the information received from the eyes, producing our sense of vision. The result is that most of what we see is actually constructed in the brain. It is your brain that sees patterns, colors, and movement in the data sent from the eyes. It is your brain—not your eyes—that picks out faces in a crowd or notices an animal running toward you.

This is the reason that programming computer vision is so difficult. It is relatively easy to outfit a robot or a self-driving car with sensors that are just as effective as human eyes—or even more effective. But it is very difficult to interpret the signals from those devices as effectively as a human brain does. A self-driving car needs to *understand* what it sees. It needs to isolate and identify the various objects within its field of vision, and to understand the potential importance of each of these objects. It needs to know whether any of these objects are moving, and at what speed and in what direction. It needs to be aware as to which of the nonmoving objects might suddenly start to move. It also needs to understand when current conditions obscure its ability to see something important, as when a parked truck blocks the view of things behind it. Effective computer vision involves a huge amount of intelligent perception, and not just sight. It also involves learning from experience, remembering what is learned, and reusing that information in appropriate situations.

A Better Model of the Principal Senses

Imagine we were all to agree on a new model of the senses. How many senses would we include in this model and what would those senses be? We could, for example, replace the five traditional human senses with the seven external human senses, as discussed earlier. However, several of our internal senses are also quite important. If we wanted to list all of the *most important* senses, what would be in the list? There is little doubt that the sense of pain is essential for our survival, and there is a good argument for including proprioception in the list. These additions would result in a model that we could call the nine primary human senses, consisting of sight, hearing, smell, taste, touch, balance, temperature, pain, and position (proprioception).

Plenty of additional senses (such as hunger and thirst) could potentially be added to the list. However, this model of nine principal human senses appears to have the strongest backing of all the alternative models. If we ever reach a consensus on replacing the five sense model, this model has a good chance of being the winner. That said, it appears unlikely we'll soon reach a consensus on this issue, which means that the traditional model of five senses will retain its dominant position for the foreseeable future.

As I was readjusting yet another misaligned roll of toilet paper, I was reminded of my earlier attempts to create a comprehensive catalog of human senses and to arrive at a definitive count of those senses. I eventually abandoned my quixotic quest, but I do feel satisfied with seven as the number of external senses, and nine as the number of primary senses.

As I rotated that roll of toilet paper into a more appropriate orientation, I began to wonder how many of my seven external senses were currently engaged. The roll was fluffy and white, quite soft to the touch, with a subtle pattern embossed on the surface. As far as I could tell, there was no added scent—thank goodness—but the roll had a faintly pleasant odor anyway. When I accidentally dropped the roll, it made a very soft sound hitting the floor. What about my sense of temperature? I noticed that the roll felt neither hot nor cold when I touched it. What about my sense of balance? I placed the roll on top of my head and stood there for a moment with my eyes closed. I didn't fall over, nor did I drop the roll of paper. What about my sense of taste? Well, no. That wouldn't be sensible. Some actions are just beyond the pale, so that one remaining sense had to go untested. But six out of seven isn't bad.

5

High Levels
of Radiation

When I'm not traveling, I seldom eat out because I cook instead. But when I travel, which I love to do, I eat out for *every* meal—preferably at a different place for each meal. My wife sometimes says, "We can just pick any restaurant that looks interesting. How about that one across the street?" In horror, I'll respond, "Oh no! We can't do that! What if the place is no good? I've got to check the ratings!" And I'll do a systematic check of the ratings of all the restaurants within a specific radius, such as 2 miles. I'll check both the Yelp ratings and the Google ratings. After eliminating the restaurants that are too pricey or don't have enough vegetables on the menu, I'll choose the three restaurants with the highest ratings. Then I'll present the results to my wife and ask her to pick from among those three—if she's still interested in eating.

My wife shows an amazing amount of patience with me. But sometimes she'll say, in a calm, patient manner, "It's not the end of the world if we don't make the perfect choice for dinner. What are you afraid of?" Of course, I know exactly what I'm afraid of. My fear is that I might make a suboptimal choice when superior choices are readily available. Furthermore, the issue of making a good choice presents me with an irresistible challenge. I love to solve optimization problems, especially when I can apply a numbers-based algorithm to address the issue.

One evening, after my wife made her choice from the three options I presented, we sat in a restaurant waiting for our food to arrive. Her words "What are you afraid of?" kept repeating in my head. I began cataloging the many things people tend to be afraid of. I soon had a very long list. In a corner of the restaurant, a TV silently displayed a news broadcast. Story after story focused on topics that reflect people's worst fears. Then a headline popped up that read "High levels of radiation in homes." I jolted with excitement. Such a wonderful addition to my list!

Soon my thoughts went off in a different direction. What does "high levels of radiation" really mean? Light is a form of radiation, yet few people cower in fear upon entering a sunny, well-lit room. On the contrary, we tend to exult in beautiful lighting and to praise the way such light makes us feel. We also surround ourselves with colorful objects, and we discern colors by discriminating among various wavelengths of *electromagnetic radiation*, which is the scientific term for light. Thus, the popular concept that radiation is harmful or dangerous is a dramatic oversimplification. A more accurate assessment is that *some types* of radiation are dangerous, while others are not. But that raises several questions. Just what does the word *radiation* really mean? What are the various types of radiation, and which ones are actually dangerous?

Origin of the Word *Radiation*

Let's start with the word *radiation*. Many people associate the term with nuclear reactors, nuclear weapons, and radioactive waste. In this interpretation, radiation is a dangerous thing. You certainly don't want to be exposed to it, and you would be terrified to discover any in your house. However, this popular use of the word *radiation* as a synonym for *nuclear radiation* (also known as *atomic radiation*) is extremely narrow because it excludes the many types of radiation that don't originate from the radioactive decay of atomic nuclei.

Before we look at the broader definition of *radiation*, let's take a closer look at nuclear radiation. You've probably heard of early-twentieth-century scientific pioneers such as Pierre and Marie Curie, who were fascinated with the newly discovered phenomenon of *radioactivity* (a term Marie coined). These scientists realized that certain substances, such as the elements radium and polonium, emitted something invisible that could affect other materials nearby— for example, causing unprocessed photographic film to fog up. This mysterious phenomenon was called radiation because it appeared to radiate in straight lines in all directions from the source substance.

For a comparison, think of a light bulb hanging in the middle of a darkened room. When the bulb is turned on, light shines in all directions, illuminating all the corners of the room. If you walk around the room, you cast a shadow, indicating that light from the bulb travels in straight lines from the source—hence the words *radiate* and *radiation*. However, we can easily see the light from a light bulb, while we can't see the radiation from a radioactive source.

Alpha, Beta, and Gamma Rays

For the early researchers into radioactive materials—such as Henri Becquerel, Ernest Rutherford, and the Curies—it was a challenge to find ways to study this invisible phenomenon. A photographic plate was used to detect the radiation, but what else could be done to yield more information? One key idea was to see if the rays could be bent by a magnetic field. These experiments soon revealed that radioactive materials emit three distinct kinds of radiation, which have vastly different abilities to penetrate other materials. It was not yet understood what these three kinds of radiation actually are, so they were simply designated by the first three letters of the Greek alphabet—*alpha*, *beta*, and *gamma* rays.

It was eventually determined that all three types of radiation result from nuclear decay. Certain kinds of atoms—the ones we call radioactive—have unstable nuclei. When an unstable nucleus decays, it transforms into a lighter kind of atom and emits radiation in the process. Alpha rays and beta rays are tiny particles of subatomic matter ejected from nuclei at high speeds, while gamma rays contain no matter at all—only energy.

Alpha particles carry a positive charge, while beta particles carry a negative charge. Alpha particles are much more massive than beta particles, and they are ejected at a much lower velocity. Because of their large size, slow speed, and electrical charge, alpha particles tend to interact with any matter in their path, which stops their progress. Consequently, alpha rays can be blocked by a sheet of paper or a few inches of air. Beta particles are much smaller and faster than alpha particles—although they also carry a charge—so it takes a lot more matter to block them, such as an aluminum panel or many feet of air. Gamma rays are extremely energetic and lack a charge, so it takes a lot of dense matter to block them, such as several inches of lead or several feet of concrete.

Realizing the Dangers

The early researchers into radioactivity were well aware that radium could cause radiation burns. Several of them—including both of the Curies—experimented by intentionally causing such burns to themselves. They could see that these burns killed skin cells and caused damage to the flesh underneath. These researchers clearly appreciated the acute (short-term) effects of high levels of exposure to certain radioactive materials. But they might not have realized the dangers associated with chronic (long-term) low-level exposure to these same materials. After all, while we all know that a hot stove can cause a serious burn, we don't assume that long-term exposure to warm objects will harm us. (If it did, snuggly house cats would be a health hazard!)

Some of these same researchers also experimented with the newly discovered phenomenon called X-rays. They noted that intense exposure to X-rays causes burns similar to the burns from radium.

Pierre Curie died in a road accident before he could experience the long-term effects of radiation exposure. Marie Curie lived much longer, to age sixty-six, and eventually died from aplastic anemia caused by the long-term exposure. Marie's daughter and son-in-law, both eminent scientists who continued her research into radioactivity, also died from ailments that were probably caused by radiation.

Unfortunately, many years passed before the dangers of chronic radiation exposure were recognized and taken seriously. In fact, in the 1910s and 1920s, people began to promote radioactive materials as healthful and curative. Again, consider the analogy with a hot stove. Although a hot stove is dangerous, sources of warmth (such a warm bath) are considered healthful. By analogy, exposure to radioactive material was assumed to be healthful provided that the dose was not strong enough to cause a burn. Many radioactive products appeared on the market, including radioactive skin creams, hair treatments, energy drinks, and pills. We now know that these products were dangerous, but it was a popular health fad at the time.

Another episode from a century ago involved the use of radium paint to create luminescent watch faces. Although the radiation from radium is mostly invisible, if you mix radium with a phosphorescent material such as zinc sulfide, the combination glows. (Radiation causes zinc sulfide to emit visible light.) Women were employed in factories to daub the hands and numerals of watches with luminous paint using very fine brushes. To keep a sharp point on the brushes, the "radium girls" were encouraged to shape the brushes with their lips and tongue repeatedly throughout the day, thereby introducing the radium into their mouths. Because radium was promoted as healthful, no one was concerned. But many of these women later suffered serious and painful health issues from the exposure, especially in their jaws, and some of them died.

The situation of the radium girls differs from the self-inflicted radiation burns of the early researchers. The researchers would place refined radium in a glass vial and then tape the vial to their skin for a few hours. The exposure was quite intense, but it ended when the vial was removed. Furthermore, the glass intercepted the alpha rays, limiting the damage to that caused by beta and gamma rays. Because the radium girls were exposed to smaller and more dilute quantities of radium, they did not experience radiation burns. However, some of that radium went directly into their mouths, allowing even the alpha rays to contribute to the tissue damage. But what made the situation so serious is that the human body easily mistakes radium for calcium, incorporating radium into the bones in place of calcium. As a result, the women suffered from continuous exposure to radiation from their own teeth and jaws.

In the case of intense short-term exposure, the immediate problem is that living cells are damaged or killed by the radiation, causing radiation burns and radiation sickness. The long-term effects are more complicated. One issue is that certain kinds of cells, such as bone marrow and white blood cells, are especially susceptible to radiation damage. That's why several of the early researchers eventually died of aplastic anemia. But today our biggest concerns are usually related to the damage harmful radiation can cause to DNA. Damaged DNA can lead to cancer and birth defects.

While we can draw valuable lessons from the unfortunate history of radium, it would be a mistake to assume that all radioactive materials present identical risks. Many different kinds of radioactive elements exist, and they differ dramatically as to the kinds of radiation they give off and how intensely they give it off. Some of these elements are toxic—without even considering the radiation effects—while others are not. Furthermore, many elements are found in both stable and radioactive forms. These different forms are called isotopes, and the radioactive forms are radioisotopes. For example, the form of cobalt used in batteries and steel is not radioactive, but the isotope cobalt-60 is a powerful emitter of gamma rays used to treat cancer. (Cancer cells tend to be far more susceptible to radiation damage than most of our normal cells.)

The Broader Meaning of *Radiation*

At the end of World War II, the United States dropped two nuclear bombs on Japan, forever changing the public perception of nuclear radiation. It became clear that such weapons present two distinct dangers: (1) the immediate danger of the intense burst of radiation released by the bomb blast, and (2) the long-term danger of nuclear fallout, the airborne radioactive waste generated by the blast. A nuclear arms race promptly ensued between the United States and the Soviet Union. By the 1960s, at the height of these tensions, many people were more concerned about the risks of fallout than the risks of the blasts—hence the popular idea of building fallout shelters.

Given the intense level of public concern, it's not surprising that the term *radiation* continued to be used as a synonym for *nuclear radiation*. Today's public concerns have largely shifted from the risks of nuclear war to the risks of nuclear power stations, especially given the catastrophic meltdowns at Chernobyl and Fukushima. In both cases, the surrounding landscape was powdered with dangerous levels of radioactive dust. And in both cases, workers responding to the emergencies were exposed to very high levels of harmful radiation. These incidents have contributed to the continued association of the word *radiation* with nuclear radiation.

Remember the key point that *nuclear radiation* means two different things: (1) ejecting subatomic particles at high speeds, and (2) emitting radiant energy without any mass. A complete definition of *radiation* should include other types of high-speed particles (not just alpha rays and beta rays), and other types of radiant energy (in addition to gamma rays). When scientists studied gamma rays to learn their secrets, they discovered that these rays travel at the speed of light.

However, only light can travel this fast. Therefore gamma rays are simply a form of light. The defining difference between gamma rays and visible light is that gamma rays have a much shorter wavelength. The upshot is that all forms of light are, in fact, radiation—and can be lumped together under the term *electromagnetic radiation*.

So now we have two different definitions of *radiation*: a narrow meaning that's synonymous with *nuclear radiation*, and a broader meaning that encompasses any kind of high-speed subatomic particle along with all forms of light. But there's a third meaning, because the word *radiation* often serves as shorthand for *ionizing radiation*. This term refers to any radiation that's powerful enough to knock electrons out of atoms or to break the bonds between atoms, thereby producing chemical changes in substances. Ionizing radiation presents a danger to living cells because it can kill the cell or damage its DNA. By some definitions, ionizing radiation consists of five principal categories: alpha, beta, and gamma rays, plus X-rays and neutron radiation. A broader definition also includes other types of particle radiation (such as protons and positrons) and the shortest wavelengths of ultraviolet light.

X-rays, like gamma rays, are a form of light. Exposure to an intense dose of X-rays can be dangerous, but it's less dangerous than a comparable exposure to gamma rays. Because the only distinction is the wavelength, and because light exists in a continuum of wavelengths, scientists have had to invent an arbitrary boundary between these two categories of invisible light. Gamma rays are those with a wavelength shorter than 10 picometers (or pm, 10 trillionths of a meter), while X-rays have waves that are longer than 10 pm.

Neutron radiation, like alpha and beta radiation, results from particles ejected from atomic nuclei at high speed. Neutron radiation is more penetrating than even gamma rays and can pass through a concrete wall. Furthermore, neutron radiation is the only form of radiation (among the five principal categories of ionizing radiation) that can cause nonradioactive materials to become radioactive. Nuclear bombs and nuclear reactors both release a great deal of neutron radiation as a result of nuclear fission, in which the nucleus of an atom splits into two separate nuclei. The resulting neutron radiation is the principal reason that nuclear bombs and nuclear power plants generate radioactive waste.

When we talk about measuring radiation, we typically focus on ionizing radiation because of the damage it can cause to human bodies. However, the issue can be approached in a few different ways. You can measure the radiation emitted by a radioactive material, expressing the results in curies or becquerels. You can measure the radiation absorbed by a person, using rads or grays as the units. However, it's important to recognize that different forms of ionizing radiation produce different amounts of damage in a human body. The solution is to use a weighted measure of the absorbed dose, reflecting the amount of potential harm. This type of measurement is expressed in rems or sieverts.

Common Sources of Ionizing Radiation

You are bombarded with radiation every moment of your life, which is fairly obvious when you consider that the broadest definition of radiation includes all forms of light. But if we consider only ionizing radiation, the fact remains that you are constantly exposed to low levels of it, called background radiation. Much of this radiation is nuclear radiation. A typical person absorbs about 300 mrem (millirems) of background radiation every year, although the amount varies from person to person and place to place. Nearly all of this radiation comes from natural sources. About 10 percent of it consists of cosmic rays, and about 75 percent comes from radon gas and other naturally occurring radioactive elements in the air. Much of the remaining 15 percent is due to the natural presence of radioactive minerals in your food and water, and in your own body.

The actual amount of background radiation you absorb varies according to your circumstances. For example, a person living in Denver (a mile above sea level) receives about twice as much cosmic radiation as someone living at sea level. If you live in an area that has a high level of uranium in the rocks and soil, you are at greater risk for high levels of radon seeping into your basement.

In the United States, in addition to background radiation, the average person absorbs 300 mrem of radiation each year due to medical imaging such as CT scans, PET scans, fluoroscopy, and X-rays (including mammograms). However, this varies dramatically from person to person. Conventional X-rays constitute only a small part of this average because the radiation absorbed from a typical X-ray is quite low.

Considering that the average person in the United States absorbs about 600 mrem of radiation each year, how close is that to being dangerous? One rule of thumb is that up to 2,000 mrem per year is considered a safe level (meaning it poses an extremely low level of risk). Another crude rule of thumb is that each 100,000 mrem absorbed increases your lifetime risk of cancer by about 1 percent. Assuming that these rules of thumb are accurate, an annual absorption of 600 mrem is nothing to worry about.

Let's now examine some specific sources of radiation in the environment, some of which present a risk and some of which do not.

COSMIC RAYS

Some of the radiation you are exposed to every day consists of cosmic rays. Cosmic rays are subatomic particles that travel through space at nearly the speed of light. Note that the term *cosmic* refers to the source of the radiation and encompasses more than one type of particle. (Likewise, the term *nuclear radiation* refers to the source of the radiation, encompassing more than one type.) However, we traditionally exclude all forms of light when we talk about cosmic rays, which limits the term to high-speed particles. By some definitions, the term *cosmic ray* includes all such particles that originate beyond the earth's atmosphere. However, some people include only particles that originate outside of our solar system. In this discussion we'll employ the broader definition, including particles that originate from the sun.

About 90 percent of cosmic rays are protons, which are in essence the nuclei of hydrogen atoms. About 90 percent of the rest are the nuclei of helium atoms (consisting of two protons and two neutrons), which are the same particles as the alpha rays emitted by radioactive decay. In other words, cosmic rays consist almost entirely of just two kinds of particles. Of the tiny remainder, some are electrons, equivalent to the beta rays emitted by radioactive decay.

Almost all cosmic rays that approach the earth are intercepted by the atmosphere before they can reach the ground. However, the high-speed collisions between cosmic rays and the molecules of the atmosphere result in a shower of other rays, some of which do reach the ground, which is why humans can be exposed. These secondary cosmic rays include X-rays, protons, alpha particles, muons, electrons, neutrinos, and neutrons.

Cosmic rays striking Earth's upper atmosphere cause some of the carbon in the atmosphere to change from carbon-12 (which is not radioactive) into carbon-14 (which is mildly radioactive). Carbon-14 is incorporated into plants when they convert CO_2 into sugar and other carbon compounds. The radioactive carbon is then passed to the animals that eat the plants and on up the food chain. Luckily, carbon-14 is not considered to be a health hazard, as your body contains a significant number of these atoms. Scientists find carbon-14 to be quite useful for dating many kinds of archeological materials, because we know the rate at which these atoms decay into other kinds of atoms. Using carbon-rich archeological materials such as wood, scientists perform carbon dating by comparing the ratio of carbon-14 atoms to other isotopes of carbon. A lower ratio indicates an older date, while a higher ratio indicates a younger date.

POTASSIUM-40

Another natural radioactive element in your body is potassium-40. Potassium occurs naturally in rocks and soil, and a small percentage of potassium atoms are radioactive. Potassium is an essential nutrient for plants, obtained from the soil. Potassium is also essential for human health. The upshot is that any food that's a rich source of potassium—such as bananas, potatoes, kidney beans, and nuts—contains a higher-than-average concentration of radioactive atoms. However, our exposure to radiation from potassium-40, like carbon-14, is so low that it presents no significant risk.

RADON

Unfortunately, not all of our natural exposure to radiation is as harmless as our exposure to carbon-14 and potassium-40. A significant risk can originate from the natural presence of uranium and thorium in rock and soil. It's not that these two elements are highly radioactive; in fact, each releases extremely low levels of radiation. Instead, the problem is that as these elements slowly decay, they produce a long list of other radioactive elements, including radium, polonium, and radon. Radon happens to be a gas, so it can seep from rocks and soil into the atmosphere. For most people, radon in the air is probably the leading source of background radiation.

If our only exposure to radon was outdoors, the risk would be quite low. However, when radon seeps from the soil into a poorly ventilated basement, the concentration of radon can greatly increase. We know that uranium miners who worked in underground mines without adequate ventilation suffered from high levels of lung cancer, primarily due to the concentrated radon gas they inhaled all day long. It's very tricky to extrapolate from this information to produce a reliable risk estimate for the levels of radon typically found in homes, but by some estimates (based on these iffy extrapolations), radon might be the leading cause of lung cancer among nonsmokers in the United States.

ELECTRONIC DEVICES

Many of the electronic devices in our homes emit small amounts of invisible radiation. This is where the ambiguity of the word *radiation* can be especially confusing. Most of these devices emit only non-ionizing radiation, which is mostly harmless and is not included when we measure harmful radiation. The only devices in our homes that emit ionizing radiation are those that emit X-rays or particle radiation. Old-style television receivers and video monitors (the ones with cathode ray tubes) often emitted small quantities of X-rays. Smoke detectors emit small amounts of particle radiation due to the presence of minute quantities of a radioisotope, as do watches and clocks with hands that glow in the dark. But even cumulatively, the exposure from these devices is usually quite minuscule—far, far less than from natural background radiation.

RADIOACTIVE WASTE

One other potential source of background radiation is the radioactive waste associated with nuclear bombs, nuclear power plants, and other sources. We often worry that these radioactive materials might get into our air, water, or food. It certainly makes sense to be vigilant on this issue and to ensure that any dangerous radioactive waste is properly handled and disposed of. The good news is that for most of us, radioactive waste is an insignificant source of background radiation—at least for now.

The Many Kinds of Light

In addition to gamma rays and X-rays, there are several other kinds of invisible light, each with its own particular range of wavelengths. These other kinds include ultraviolet light, infrared light, microwaves, and radio waves. The sun emits all of these wavelengths—in addition to visible light—and they all travel from the sun to the earth at the speed of light. However, these wavelengths are not emitted in equal amounts, and not all of them are capable of penetrating Earth's atmosphere to reach the ground.

Just as a rainbow includes all wavelengths of visible light, the full spectrum of electromagnetic radiation encompasses all forms of light. This spectrum continues on beyond red at one end of the rainbow and violet at the other end of the rainbow. As you go beyond red, through

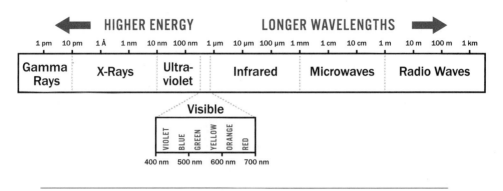

The complete spectrum of light (electromagnetic radiation)

infrared and on to microwaves and radio waves, the wavelengths get longer. As you go beyond violet, through ultraviolet and on to X-rays and gamma rays, the wavelengths get shorter.

The longer the wavelength, the less energy each photon of light contains. This may seem odd because we think of microwaves as having a lot of energy, when in fact a microwave photon has less energy than a photon of visible light. It's the other end of the spectrum, starting in the ultraviolet range and continuing on to X-rays and gamma rays, where the photons of light have enough energy to damage your cells. As the wavelengths get shorter and shorter, the radiation tends to be more dangerous to living things, although the correlation is far from perfect. (The actual risk depends not only on the wavelength but also on several other factors, including the intensity and duration of the exposure.) As you can see from the diagram, gamma waves have extremely short wavelengths and hence tend to be particularly dangerous to us.

ULTRAVIOLET LIGHT

Ultraviolet light, often called UV for short, has wavelengths that are intermediate between X-rays and visible light. Consequently, the energy in UV rays is intermediate between X-rays and visible light. Because the spectrum of light is continuous, scientists had to invent an arbitrary dividing line between X-rays and UV light, and they placed that boundary at 10 nanometers (or nm, 10 billionths of a meter).

Wavelengths shorter than 10 nm are called X-rays, and wavelengths longer than 10 nm are called ultraviolet.

The UV band of wavelengths can be divided into narrower bands, called UV-A, UV-B, and UV-C. This can be helpful for several reasons. First, human skin reacts differently to the different bands of UV and to different areas within each band. Second, some of the UV-absorbing compounds in sunscreen formulations can block UV-A or UV-B but not always both. And third, Earth's atmosphere selectively blocks certain wavelengths of UV light. UV-A passes through the atmosphere easily, while UV-B is partially blocked, and UV-C is completely blocked.

The wavelengths of UV that reach the earth are not short enough (and therefore not strong enough) to ionize molecules, so UV-A and UV-B are not classified as ionizing radiation. However, UV light is powerful enough to cause other types of chemical reactions that can change some of the molecules in your skin. This can result in visible aging of the skin, and damage to DNA molecules can cause skin cancer. Although we don't include UV light when we calculate radiation levels, we all know that intense exposure to UV light from the sun can cause sunburn. (We don't normally call this a radiation burn, but in a sense it really is.) Such a burn may result in the skin blistering or peeling away, indicating that many skin cells were killed by the exposure.

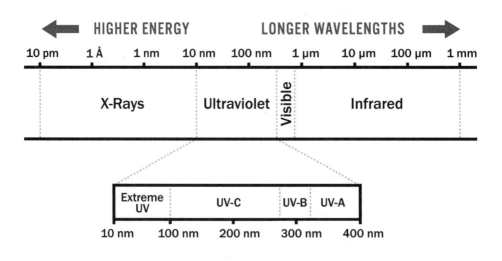

Subdivisions of UV light

Because of the damage UV light can cause to eyes and skin, it's important to be cautious with devices that emit UV, especially "black lights" that might not emit any visible light at all. Most black lights emit only UV-A and UV-B, but tubes can be purchased that emit UV-C instead. UV-C tubes can be quite hazardous and should be used only with protective goggles and clothing.

VISIBLE LIGHT

The range of wavelengths humans can see is called visible light. When we see a rainbow, it's because drops of water in the air have divided sunlight into its component wavelengths. One end of the visible spectrum appears red, and the other end appears violet, with several other colors in between. The exact boundaries of the visible spectrum are a bit fuzzy because our ability to see the light fades away at either end rather than cutting off abruptly. Defined in round numbers, we might say visible light runs from wavelengths of about 400 nm to about 700 nm.

The various wavelengths of visible light are selectively absorbed by the materials around us, including our skin. Only certain wavelengths bounce off each object to reach our eyes, which is why the materials appear to have different colors. When visible light is absorbed, the energy does not disappear but instead is transformed into other forms of energy—usually heat.

Visible light is a non-ionizing form of radiation, and until recently, most people assumed that visible light is usually harmless, although we know not to look directly at the sun because the light is so intense. However, the internet is now rife with rumors about the dangers of blue light. The original science behind this idea—the precursor to the rumors—suggested that using electronic screens late into the night might disrupt your sleep patterns and that reducing the amount of blue light from these screens in the hours before bedtime might improve your sleep. From this modest beginning, articles in the popular media now claim that blue light can do serious harm to your eyes or skin, and that phones and computer screens are the main source of this danger. Very little evidence exists to support these extreme claims.

INFRARED LIGHT

Infrared (IR) light consists of longer wavelengths than visible light and thus has lower energy per photon. Although IR light is invisible, we often notice its effects because of the huge role it plays in the transfer of energy between objects. Whenever we feel radiant heat coming from something, such as a space heater or radiator, we're actually feeling the IR light emitted by the object. In fact, all objects constantly emit infrared radiation, but we only notice the strongest emitters. Admittedly, this is a counterintuitive concept. If an object does not visibly glow, we have a hard time thinking of it as emitting light.

IR emissions are strongly tied to the temperature of the object. The warmer an object is, the greater the amount of energy it radiates and the shorter the average wavelength of the light it gives off. If an object gets hot enough—around 977 degrees Fahrenheit (525 degrees Celsius)—a tiny fraction of the emitted light shifts into the red end of the visible spectrum, just enough to be seen as a faint red glow. For example, if you start up a burner on an electric stove, you can soon feel energy radiating from the burner. Eventually the coil becomes hot enough to glow a faint red.

Most materials (including skin) are quite good at absorbing IR light, although there are important exceptions such as aluminum foil and snow. If you sit close to a fireplace while a log is burning, the parts of your body and clothing that face the fire soon become quite warm. This is because IR light as it is absorbed is converted to heat. Everything around you (yourself included) simultaneously gives off IR and absorbs IR, exchanging energy with all of the surrounding objects. Any object that gives off more than it absorbs loses energy and becomes colder, and anything that absorbs more than it gives off gains energy and becomes warmer. All the objects around us also exchange energy due to heat conduction with the things they touch, so separating out the two effects can sometimes be tricky.

Note that although the word *radiation* is part of the term *infrared radiation*, IR is not a type of ionizing radiation. IR will not damage living tissue unless you absorb so much of it that the rising temperature of your skin causes a burn. But in such a case, the absorption of IR photons is not the direct cause of any damage; the excessive buildup of heat causes the damage. In general, we don't usually fear infrared radiation, and in fact, when temperatures are low we relish feeling the thermal radiation from warm objects.

MICROWAVES

As we continue our journey through the spectrum of light, past the range of infrared radiation, we reach the zone of microwaves. Again, because the spectrum of light is continuous, scientists have had to create an arbitrary dividing line between infrared and microwaves, and another line between microwaves and radio waves. The most common definition is that microwaves have a wavelength between one millimeter and one meter. It may seem ironic to call these waves micro, since they're longer than all other forms of light except radio waves. However, some people consider microwaves to be a subcategory of radio waves, and when categorized this way, microwaves occupy the short-wavelength end of the radio spectrum, which is how they got the name.

Until the early 1970s, microwaves were best known for their role in our communications infrastructure, primarily in the form of microwave relay towers. Now the term *microwave* is strongly associated with a kitchen appliance—a small oven that employs microwaves to heat food and beverages. Before these ovens became popular, the public seldom thought about microwaves. These appliances introduced a mysterious new form of energy into our daily lives, leading to a lot of speculation and rumor, such as the myth that microwaved food is dangerous to human health.

Microwaves cause food and water to heat up for essentially the same reason that a fire in the fireplace heats up your body and clothing: because invisible light is absorbed and converted to heat. However, there is one important difference. Although most materials are good at absorbing infrared, far fewer substances are good at absorbing microwaves. Water is especially good at absorbing microwaves of a specific length, around 12 cm long, so microwave ovens are tuned to emit this particular wavelength. When we heat food in a microwave oven, most of the energy is absorbed by water molecules, which share the heat with neighboring molecules. Fats, sugars, and certain other food molecules also absorb microwaves—although less efficiently than water—so these other molecules also contribute to the warming.

Because microwaves are longer than IR waves, and much longer than visible light, they contain less energy per photon. Therefore microwaves are not a form of ionizing radiation, and they won't directly cause any damage to cells or DNA. As with infrared radiation, any damage to flesh would be indirect, as a result of the rise in temperature. The fact that microwave ovens are designed to keep the radiation inside the oven reinforces the popular fear that such radiation would be dangerous if it escaped. However, our daily lives are filled with devices that send or receive signals via microwaves: routers, smart speakers, laptops and smartphones with Bluetooth and Wi-Fi capabilities, Bluetooth headsets and baby monitors, GPS devices, garage door openers, and satellite TV, to name just a few. Most of these devices operate at very low power compared to a microwave oven, so they would be useless for reheating your cold cup of coffee. (Not to mention that they aren't tuned to the right wavelength!)

RADIO WAVES

Radio waves are extremely long—ranging from a meter in length to more than a kilometer—and therefore they have less energy per photon than any other form of light. However, shorter radio waves easily pass through the atmosphere and other materials, which make them great for communication. We can embed information in radio waves by modulating the signal (making rapid changes to the amplitude or the frequency), and this is how we broadcast radio and television. We

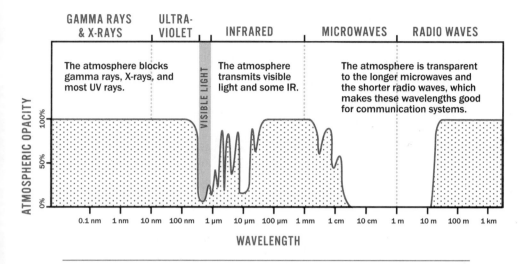

Atmospheric opacity to different wavelengths of light

divide the radio wave spectrum into various frequencies (which are inversely related to wavelengths) and assign these different frequencies to different uses. In addition to radio and television broadcasts, we also use radio waves for aircraft communication, marine communication, amateur radio, and RFID chips.

Radio waves are far too long for the photons to ionize matter, and the longer radio waves tend not to be absorbed by matter at all. Thus it's unlikely that radio waves cause any direct harm to living tissue or any molecular changes that might lead to cancer. However, there is still a slight possibility that certain wavelengths of radio waves in very high doses could have a harmful indirect effect due to a completely different mechanism, yet to be identified. So far, no convincing scientific evidence of any such danger has emerged, which means that we currently have no good reason to fear the radiation emitted by cell phones, smart meters, 5G cell towers, or any of the many other devices that communicate by emitting radio waves or microwaves. It is also worth noting that most of the devices used in homes emit radiation at a very low intensity, which is why they can operate for hours powered only by tiny batteries. In fact, a human body gives off far more radiation (mostly in the IR band) than most of these devices.

What Were Those Categories Again?

It must be apparent to you by now—if it wasn't already—that the word *radiation* refers to quite a range of phenomena. In this chapter I've mentioned many kinds of particle radiation plus seven categories of electromagnetic radiation (visible and invisible light). I mentioned that the terms *nuclear radiation* and *cosmic rays* each refer to radiation from particular sources, but that both terms encompass more than one type of radiation. And I mentioned that the term *ionizing radiation* refers to any type of radiation that carries enough energy to detach electrons from atoms or molecules, thereby ionizing them. (As a result, ionizing radiation can be dangerous in large doses, while most non-ionizing radiation is harmless.) The following table may be helpful in sorting this all out.

While it makes perfect sense to divide these types of radiation into those that are dangerous and those that are not, we should not forget that this is a simplistic model. To assess the risk of radiation exposure, we must consider not only the type of radiation but also the intensity

A QUICK REVIEW OF ELEVEN TYPES OF RADIATION

	SUBATOMIC PARTICLES	ELECTROMAGNETIC RADIATION	IONIZING RADIATION
ALPHA RAYS	●		●
BETA RAYS	●		●
NEUTRON RADIATION	●		●
PROTON RADIATION	●		●
GAMMA RAYS		●	●
X-RAYS		●	●
ULTRAVIOLET LIGHT		●	
VISIBLE LIGHT		●	
INFRARED LIGHT		●	
MICROWAVES		●	
RADIO WAVES		●	

and duration of the exposure. But to truly understand the risks, scientists go a step farther, identifying exactly how different wavelengths of light and different types of particle radiation interact with the various molecules of a human body—and what molecular changes can occur as a result. Scientists then evaluate how these modified molecules can affect the normal functioning of human cells and tissue. This analysis produces the most convincing evidence of whether a particular type of radiation poses a risk, and how much exposure is necessary to generate the risk.

In the restaurant where I'm sitting with my wife, the food has still not arrived, although our server assures us it will be right out. I look around the room, thinking about the various kinds of invisible yet harmless radiation. A message pops up on my cell phone, having arrived via microwaves sent out from a nearby cell tower. I think I hear the beep of a microwave oven in the kitchen, perhaps reheating something that will soon appear on my table. It's a cold evening outside, but a space heater placed near our table is bathing us in infrared radiation, keeping us cozy.

But it's the visible radiation I most appreciate—the carefully placed lights and the colorful objects that adorn the room. It's my ability to detect this visible radiation that gives the restaurant its pleasant ambiance. I think back to where we ate lunch earlier in the day, a place full of light streaming in through large glass windows. I really enjoyed sitting there; it made me feel good. So while it may be scary to read news articles or see TV headlines about radiation, I conclude that I really appreciate most of the radiation that surrounds me.

I pull out my pen to start a list of the many ways radiation makes my life more pleasant, but then the food arrives. So I decide to eat instead.

6

Killing Germs

I have a guilty little habit that I'm reluctant to admit. At the risk of lowering your opinion of me, I'll just come out and say it: I often eat yogurt directly from the container instead of putting it into a bowl and then eating it. You're probably picturing one of those tiny single-serve containers that holds a few ounces. No, I'm talking about those really big containers, the ones that hold 24 or 32 ounces of yogurt. Once upon a time, I would put yogurt into a bowl before eating it, but now I usually don't. If I'm the only person in the house eating this yogurt, it shouldn't matter, should it? No one else is going to catch my germs. And yet I feel a bit guilty, as if I'm doing something naughty or unsavory.

I seldom eat a lot of yogurt all at once. I eat a few spoonfuls and then put the container back into the refrigerator to enjoy again the next day. My usual time for such a snack is just before midnight. As I eat, I don't dip my spoon deeply into the yogurt but instead glide it sideways, always maintaining a smooth surface on the yogurt. I do this for aesthetic reasons; a smooth surface looks so much better than an uneven surface. No matter how much of the yogurt I've consumed, I make sure the surface remains glassy smooth.

So there I was standing in the kitchen late one evening, enjoying my ration of yogurt, when my thoughts turned to the topic of germs. Does my yogurt-eating method deposit fewer germs than a choppier method would? What kind of germs, if any, was I leaving behind?

Would those germs actually survive in the acidic yogurt? As I pondered such questions, my eyes fell upon the ingredient list, which included seven different species of live bacteria. Clearly these seven species are not germs—or are they? What does the word *germ* really mean? Does the word *germ* refer only to harmful microorganisms, or are there good germs too? After all, we constantly talk about killing germs, which suggests that germs are bad.

The idea of killing germs is deeply entrenched in the way we think about the world. Parents repeatedly warn their children about the danger of germs. As adults, we're exposed to a barrage of advertising about germs, offering an endless assortment of products to kill germs on our hands, in our mouths, on our kitchen countertops, and in our toilet bowls. Even before COVID-19, this obsession generated a lot of fear—such as the fear of touching toilet handles or bathroom doors in public buildings—and it fueled a lot of demand for antimicrobial products. Similar fears and precautions have long caused a demand for antibiotic medicines (which can treat bacterial diseases but not viral diseases). But now we hear in the news that overuse of antibiotics has caused some very serious problems, particularly the rise of antibiotic-resistant strains of bacteria. How do we sort this all out?

The Concept of Germs

Although the concept of germs is exceedingly familiar to us now, it was not until the invention of the microscope that people became aware of their existence. Antonie van Leeuwenhoek, a pioneer builder of microscopes, stunned the scientific world in the 1670s with his discovery of microscopic organisms. In fact, he found a hugely diverse world of little creatures too small to see with the naked eye. Everywhere he looked with his microscopes, he found still more of these mysterious little beasts.

The connection between this amazing discovery and disease was not immediately recognized. As late as the 1850s, most scientists still did not believe that microorganisms caused disease. Being scientists, they didn't attribute diseases to witchcraft or evil spirits, as many people did, but instead followed the clues. Some scientists noticed that after certain diseases appeared somewhere, they would often spread. This suggested some sort of contaminating agent, either clinging to objects or else floating in the water or the air. But most scientists at the time assumed that the contaminants were nonliving pollutants. For example, malaria was assumed to be caused by bad air—hence the word *malaria*, which literally means "caused by bad air."

A minority view among scientists was that these contaminants consisted of tiny particles of living matter, like microscopic seeds. The idea was that these tiny seeds, or germs, could come into contact with a person and then sprout into a disease. When scientists in the mid-1800s began to take this idea more seriously, it became known as the germ theory of disease. The word *germ* has stuck with us, even though we no longer think of germs as seeds. In other contexts, we still use the word with the original meaning. For example, a sprouting seed is said to germinate. Wheat germ is the part of the wheat seed that sprouts. And if you come up with a germ of an idea, it doesn't mean your thinking is diseased; it means the idea is just beginning to sprout in your mind.

Still, the core idea of germ theory—that diseases are often caused by microscopic living things—turned out to be correct. Once this theory was widely accepted, it transformed our thinking about how to avoid infectious diseases and how to treat the diseases we fail to avoid. In the popular imagination, the principal solution to both matters—avoidance and treatment—is to kill germs. But the issue is actually more nuanced than that, in part because germs are an amazingly diverse set of creatures. Furthermore, we now know that certain microorganisms are actually quite beneficial and that their indiscriminate slaughter is not in our own best interest.

So What Exactly Are Germs?

Although scientists invented the word *germ*, it's no longer used in scientific discussions, in part due to its vagueness. The word remains popular with the general public, who primarily associate it with the spread of disease by bacteria and viruses. Thus, the narrowest definition of the word *germs* would be "bacteria and viruses that cause human diseases." Yet most people have a broader concept of germs, implying a definition that includes one or more of the following additional categories:

- bacteria and viruses that cause diseases in animals (not just in humans)
- bacteria and viruses that cause diseases in plants
- any microscopic thing of biological origin that can cause disease, including protists (formerly called protozoa), fungi, and prions
- all bacteria, including harmless and beneficial species (the so-called good germs)

The upshot is that the word *germ* effectively conveys a vague general concept, but the term is useless for anything more precise.

A related word, somewhat more precise, is *microorganism*. This word refers to any living creature that's too small to see with the naked eye. This term covers all bacteria, all protists (single-celled creatures that are more complex than bacteria), all microscopic fungi, and a few other categories. Because viruses are not actually living creatures (by any strict definition, as we shall see shortly), ambiguity exists as to whether viruses should be called microorganisms. Some sources say that viruses are indeed microorganisms, while other sources say no. By the way, a short and somewhat informal synonym for *microorganism* is *microbe*—a much easier word to spell!

A microorganism that causes a disease is called a pathogen. In fact, the definition of *pathogen* includes any microscopic thing of biological origin that causes disease, so viruses are also included. (Viruses are definitely biological, even if they're not alive.) Calling something a pathogen tells us that a particular microscopic thing is *capable* of causing a disease, even if it seldom does. For example, several species of microorganisms that are normally harmless to humans can cause disease in people whose immune systems are weakened.

BACTERIA

When we talk about killing germs, the targets of our fatal intentions are usually bacteria, which are extremely tiny living creatures. Some of the human diseases caused by bacteria include cholera, tuberculosis, typhoid, tetanus, Lyme disease, chlamydia, salmonellosis, syphilis, diphtheria, leprosy, bubonic plague, pertussis (whooping cough), listeriosis, psittacosis, rheumatic fever, scarlet fever, anthrax, and strep throat. In addition to causing these human illnesses, bacteria cause a wide range of diseases in other creatures, including plants. But most types of bacteria don't cause disease at all, instead living out their lives in ways that are harmless to us, or even helpful.

Bacteria have been on Earth for several billion years, much longer than almost any other form of life. (The one exception is the Archaea, another category of single-celled creatures, almost as ancient as bacteria.) We don't know how many species of bacteria currently exist, but it is a huge number—probably hundreds of thousands of species, possibly even millions. In addition to their enormous variety, bacteria are also amazingly numerous. A single spoonful of soil typically contains millions of bacterial cells. Your body harbors trillions of bacteria, some of which are on your skin and some of which are inside your body. The idea of killing off all of these bacteria—well, it just isn't going to happen, no matter how hard you try.

But because bacteria are alive, they can indeed be killed, even if it's impossible to kill off every last one. One way of killing bacteria is to subject them to temperatures too high or low for them to survive. That's why we're advised to heat raw meat to a certain temperature when cooking it. However, different kinds of bacteria have different tolerances for high or low temperatures, so cooking isn't guaranteed

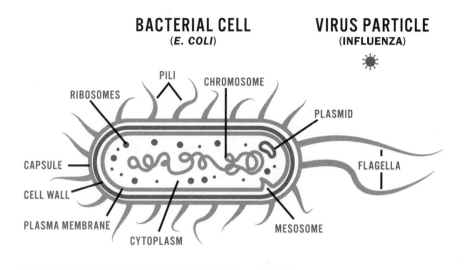

BACTERIAL CELL
(E. COLI)

VIRUS PARTICLE
(INFLUENZA)

RIBOSOMES

PILI

CHROMOSOME

PLASMID

CAPSULE

FLAGELLA

CELL WALL

PLASMA MEMBRANE

CYTOPLASM

MESOSOME

Cross section of a bacterium, and size comparison with a virus

to kill 100 percent of all the kinds of bacteria, even though it should make the meat a lot safer to eat. On the other hand, when you get sick, a fever in your body can be warm enough to make life difficult for certain kinds of bacteria.

Another way of killing bacteria is to expose them to toxic chemicals. The tricky part is to identify chemicals that are highly toxic to bacteria while doing little or no harm to the cells of your body. This is especially important if the plan is to swallow or inject the material, attacking bacteria that are inside your body. Whenever we find or invent such a chemical, we can add it to our small arsenal of antibiotics, medicines specifically intended to kill internal bacteria. The problem is that bacteria can become resistant to our new antibiotics, sometimes very quickly, a topic we'll explore in a few moments.

Each bacterium consists of a single cell, while a human body consists of trillions of cells. The cells of bacteria are typically much smaller and much simpler than the cells of humans or any other kind of life. We call them cells because each bacterium is surrounded by a membrane that holds the bacterium together and separates it from the rest of the world. Furthermore, the membrane regulates the materials that pass into and out of the cell. Inside the membrane is a gooey mixture of water, proteins, fats, carbohydrates, DNA, and other substances.

This cytoplasm is actually quite similar to the cytoplasm inside our own cells, which is part of the challenge in finding new antibiotics that effectively kill bacteria without harming our own cells.

Because bacteria are alive, they do several things that many other living things also do. They "eat" by absorbing external materials that contain essential compounds. They "breathe" by absorbing the gases they need and giving off waste gases. When they have access to food, they grow and become larger, and they reproduce by splitting in two. Many bacteria can also swim, using either a whiplike flagellum or hairlike cilia to propel themselves. Bacteria often exchange DNA with other bacteria, resulting in new genetic combinations that help the bacteria to survive and adapt to changing conditions.

Being alive, bacteria have a metabolism. In other words, various biochemical processes constantly happen within the cell, and these processes consume energy. If the cell runs out of energy, and if the cell cannot replace the energy by consuming an appropriate food, the cell will die. Each of the metabolic processes in a bacterium provides us with a potential avenue of attack for a new antibiotic. We just have to find a chemical that will interrupt a key metabolic process in a bacterial cell without interrupting any similar processes in a human cell.

VIRUSES

Other than bacteria, the other familiar category of germs is viruses. Because we call both groups germs, many people assume that the two groups are fairly similar and that whatever "kills germs" will also kill viruses. In fact, bacteria and viruses are dramatically different things. Furthermore, nearly all of our antibiotics are completely ineffective against viruses. You cannot cure a cold (which is a viral disease) by taking an antibiotic.

As with bacteria, there are many different kinds of viruses, infecting not only humans but also animals, plants, and even bacteria. Human diseases caused by viruses include the common cold, influenza (flu), COVID-19, chickenpox, HIV/AIDS, herpes, mumps, measles, German measles (rubella), shingles, viral hepatitis (types A through E),

Zika, chikungunya, rabies, polio, West Nile, dengue, yellow fever, SARS, MERS, and Ebola.

A virus is much smaller and simpler than a bacterium, which is much smaller and simpler than a human cell. In fact, viruses are so small they were not discovered until the 1890s, long after the germ theory of disease had become widely known. A virus is typically just a fragment of RNA or DNA enclosed in a protective wrapper. (Some viruses contain a bit more than this, but not much more.) A virus has no cytoplasm. It has no permeable membrane, allowing selected materials to go in and out. It has no metabolic processes. It does not consume food. It does not exchange gases with its environment. It does not expel waste. It cannot starve. It cannot swim or move. Unlike a bacterium, a virus cannot reproduce on its own. In fact, until a virus bumps into an appropriate host cell, it remains a completely inert particle, without any of the essential features we associate with living things, except that it contains a fragment of DNA or RNA.

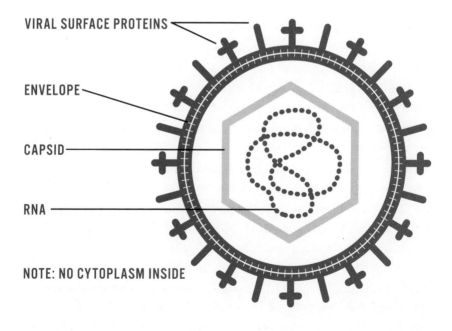

Anatomy of a typical virus

What a virus does have are two very important features that make it extremely powerful:

1. The surface proteins of a virus can adhere to the cell membrane of an appropriate host cell, after which the viral RNA or DNA enters the cell.
2. The invading RNA or DNA redirects the metabolic activities of the hijacked cell, turning the cell into a factory to crank out lots more virus particles.

Thus, viruses are not just random bits of genetic material; they are bits of genetic material that are capable of hijacking living cells. The virus does not need to contain all the genetic information necessary to run the hijacked cell. It needs only enough RNA or DNA to redirect the activities of the cell. This can be compared to modern-day pirates who hijack an oil tanker. The pirates might arrive next to the giant tanker in a tiny boat, even a rubber dinghy. Once aboard the oil tanker, the hijackers don't need to know all the details of how to run the ship; they simply need to coerce the captain and crew to follow their orders. Likewise, the RNA or DNA in a virus takes over the hijacked cell, but some of the cell's original DNA is still needed to keep the cell alive and operating.

With the machinery of the hijacked cell redirected to manufacturing more virus particles, the newly created viruses need a way to escape in order to infect other cells. In some viral diseases, no viruses escape the host cell until the cell is packed full of new virus particles, at which time the cell ruptures, killing the cell but releasing a great quantity of the virus. In other viral diseases the new viruses can escape by budding off while the cell continues to manufacture more virus particles. Note that the only way for a virus to reproduce—or to do anything at all—is to hijack a living cell from a susceptible species.

Because all viruses make their living by invading living cells, you might say that all viruses are harmful. However, many types of viruses are completely harmless to humans, even though they harm certain other species of life. Each type of virus requires a particular host species—or range of related host species—in order to infect and kidnap the corresponding cells. For example, some viruses attack only bacteria. We call such a virus a *phage*, short for *bacteriophage*. Now imagine a phage that attacks a species of bacteria that causes human illness. The presence of the virus could actually protect you from the bacterial disease. Such viruses actually exist, and scientists are becoming more aware of their potential value. In the future, instead of relying solely on antibiotics to fight bacterial diseases, we may increasingly turn to phage therapy as a way of treating certain diseases.

If a virus is not alive, and by the strictest definitions it's not, does it make any sense to speak of killing a virus? Even if a virus cannot literally be killed, it can certainly be destroyed, which has a similar result. You could argue that it makes perfect sense to speak of killing a virus, especially with regard to destroying virus particles on your hands and on surfaces around you. However, after a virus particle has entered your body, it is often quite difficult to destroy it. Until it hijacks a cell, such a particle has no metabolism, which limits our options for attacking it. And after the particle hijacks a cell, it becomes a part of the cell, and "killing" the virus would mean killing your own cell. The upshot is that it is much harder to discover highly effective antiviral drugs than to discover effective antibiotics. Therefore, with most viruses, our principal line of defense is to prevent the disease (through vaccines and good hygiene) rather than curing it.

On the other hand, we have made significant progress in recent years in identifying new antiviral drugs. In contrast to antibiotics, which are used to kill bacteria and can frequently cure bacterial diseases, a typical antiviral does not attack the virus directly and does not cure the disease. Instead, most antivirals slow down the reproduction of viruses within your body, which means you'll be less sick while you suffer from the disease. For any disease that can be fatal, this improves your chance of survival. Because of the reduced viral load in your body, your body's defenses can be more effective in fighting back, sometimes reducing the length of the illness. Antivirals have been especially successful in fighting HIV/AIDS, allowing people with the disease to live long and relatively normal lives, even though the treatment does not cure them of the disease. Another notable and quite recent success is the antiviral treatment for hepatitis C, which can actually clear the virus from the bodies of most patients.

OTHER TYPES OF PATHOGENS

When someone mentions germs, we usually think only of bacteria and viruses, but other pathogens could also be called germs. Most of these pathogens are living creatures with cells that are more complex than those of bacteria, although certain pathogens are even smaller and simpler than viruses.

Living creatures with complex cells that include a distinct nucleus are called eukaryotes. All of the plants, animals, and fungi we can see with our naked eyes are multicellular eukaryotes. However, some eukaryotes are microscopic single-celled organisms. Today we refer to all single-celled eukaryotes as *protists*, although we used to call many of them *protozoa*, a term used less often now. Diseases caused by protists include malaria, amoebic dysentery, giardiasis, toxoplasmosis, cryptosporidiosis, Chagas disease, leishmaniasis, and African sleeping sickness. But again, as with bacteria, the vast majority of protist species are harmless to us. Although some people refer to disease-causing protists as germs, it is more common to call them parasites. The word *parasite* also applies to multicellular organisms that can invade human or animal bodies, such as tapeworms and hookworms.

And speaking of hooks, we shouldn't let fungi off the hook either. Several human diseases are caused by various kinds of fungi, including nail infections, yeast infections (candida, thrush), athlete's foot (ringworm), blastomycosis, and histoplasmosis. Interestingly, a fungal spore is something like a microscopic seed, very similar to the original concept of a germ. We don't often think about fungal diseases when we talk about germs, but hygienic measures such as hand washing are helpful in reducing exposure to such diseases, just as with bacteria and viruses.

At the tiny end of the size spectrum are several kinds of viroids that can infect plants. *Viroids* are like small viruses, except that they

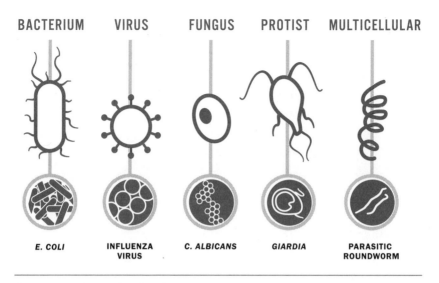

BACTERIUM	VIRUS	FUNGUS	PROTIST	MULTICELLULAR
E. COLI	INFLUENZA VIRUS	C. ALBICANS	GIARDIA	PARASITIC ROUNDWORM

Categories and examples of pathogens

lack the protein coat. Even smaller are prions, which are misfolded protein molecules. The odd thing about prions is that they can cause normal proteins in the brain to become misfolded, thereby causing a chain reaction that destroys the functioning of the brain tissue. The best-known examples of prion diseases are mad cow disease and the closely related Creutzfeldt-Jakob disease.

Immunization and Prevention

It's certainly better to avoid catching a disease than to catch it and hope it can be cured. Many infectious diseases, especially viral diseases, have no cure. For example, influenza (flu) and the common cold, both viral diseases, have no cure. If you get sick with the flu or a cold, you just have to wait it out until you get better, although certain over-the-counter medicines can help you feel a bit better while you wait, and certain antivirals can sometimes decrease the severity of the disease.

Frequent and thorough hand washing has long been emphasized as a particularly helpful technique to reduce the spread of both viral and bacterial diseases. If you are the person carrying the germs, hand washing reduces the odds of your passing the disease to someone else. If you're healthy, washing your hands greatly reduces the odds of picking up a disease from the soil or from another person. If soap and water aren't available, hand sanitizer can help reduce germs, although the CDC (Centers for Disease Control) says this method is not as effective as using soap and water.

Of course, dirty hands are not the only way diseases are spread. Intensive research into the spread of COVID-19 has revealed that this disease is almost always contracted by breathing contaminated air rather than by touching contaminated surfaces. Contaminated water or food can also be a source of germs. There are techniques to reduce the spread of disease via each of these routes. The word *hygiene* is a shorthand way to refer to this collection of techniques and practices. On the other hand, quite a few serious diseases, including malaria, are spread by mosquitos or other pests, in which case the principal focus is to reduce the odds of people being bitten by the infected creatures.

In the modern world, we also rely on immunization, in addition to hygiene, to reduce the risk of becoming infected by a disease. Immunization, also called vaccination, is a way of training your body's immune system to recognize pathogens quickly so your body can mount a response before the germs can make you sick. The vaccine you're given may contain weakened germs, inactivated ("dead") germs, fragmented parts of germs, or substances that mimic the surface of germs. Most of these methods allow your body to become familiar with the surface proteins of the pathogen and thereby to create antibodies to attack it. It's a bit like putting a used article of clothing under the nose of a bloodhound so that it knows what scent to search for. Later on, if you should be exposed to live germs of the same type, your body is ready to launch a rapid counterattack, disabling the germs before they can disable you.

For a few diseases, such as tetanus and diphtheria, the vaccine trains your body to recognize the toxins produced by dangerous germs rather than recognizing the germs themselves. This allows your body to quickly neutralize the deadly toxins produced by these particular germs.

New types of vaccines involving viral RNA have recently become quite important, playing a leading role in the race to create COVID-19 vaccines. This type of vaccine is really just another way of introducing your body to a "spike protein" from the surface of the virus. But instead of actual germs or proteins, the vaccine contains messenger RNA (mRNA). This molecule contains coded information that tells a cell how to make the protein. Immune cells in your blood gobble up the injected mRNA, which causes some of these cells to produce the protein and display it on their surfaces. This trains your body to recognize the protein, thereby generating the immune response. The advantage of this approach is that new vaccines can be developed more quickly and manufactured more easily. It also eliminates the need to culture the germs (which can be slow and difficult) or to inject anybody with the germs.

Vaccination programs have been in widespread use since the 1880s, and as time has gone by, vaccines have been developed for more and more diseases. These programs have saved tens of millions of lives. Some once-feared diseases that used to be quite common, killing or harming many thousands of people each year, are now quite rare. One deadly disease, smallpox, has been completely eradicated, primarily through worldwide vaccination programs. However, for most diseases, we have not eliminated the germs—we have simply made people immune to the germs. If we fail to continue vaccinating people for these diseases, the diseases will soon return.

Unfortunately, the internet is rife with false attacks on the safety and efficacy of vaccines, including erroneous claims that vaccines cause autism, that vaccines contain dangerous ingredients that are likely to cause harm, and that vaccines don't actually work. On the contrary, approved vaccines tend to be quite safe and effective—far, far safer than letting the corresponding diseases run through the population. Vaccines not only protect the immunized individuals but also provide us with a safe way to achieve herd immunity. In other words, when nearly all of the people in a particular place have been vaccinated, the disease is highly unlikely to spread through the community even if a carrier starts spreading the germs.

Overuse of Antibiotics

Use of antibiotics to treat bacterial diseases is a relatively recent development. Sulfa drugs (our oldest category of antibiotics) began to be used in the late 1930s, and penicillin went into production as a medicine to treat disease in 1942. In other words, our reliance on antibiotics is less than a century old. In this brief period of time, science has developed several distinct classes of antibiotics, and each class contains several different drugs. Antibiotics quickly proved to be miracle drugs, genuinely deserving of that overused praise. In case after case, a person with an extremely serious bacterial infection that would probably have been fatal was quickly cured through the use of antibiotics. The discovery of antibiotics is one of the greatest achievements of twentieth-century medicine, and their use is credited with saving millions of lives.

However, that achievement is rapidly being undone. It turns out that bacteria quickly become resistant to any antibiotic that's heavily used. As we turn from one antibiotic to another in an effort to kill germs, some strains of bacteria become resistant to all of them. These multidrug-resistant bacteria pose a serious threat to human life because we have few good ways left to kill them. A noteworthy example is a resistant form of staph infection (*Staphylococcus aureus*) that tends to spread in hospitals and other institutions that treat patients. Many other types of resistant bacteria have also developed, resulting in a growing list of diseases that can be difficult to treat.

Bacteria become resistant because whenever antibiotics are used, they seldom kill off all of the harmful bacteria, even though they might kill off enough to cure the patient. The few bacteria that survive are the ones that are most resistant to the drug. These survivors pass on their resistance to their offspring. Furthermore, because of the way bacteria exchange DNA with each other, a resistant bacterium can pass its resistance to other bacteria that are only distantly related. In other words, a resistance that develops in one species of bacteria can be transferred to another species of bacteria. Thus, every use of an antibiotic helps to breed resistance to that antibiotic.

The logical solution is to greatly reduce our use of antibiotics, employing them only if we strongly suspect a dangerous bacterial infection. We should not insist on being given an antibiotic every time we catch a cold, in part because a cold is a viral infection that cannot be cured with an antibiotic and in part because a cold is a minor disease that does not present a serious risk to most people. But overuse of antibiotics is not limited to the doses given to people. Antibiotics are routinely given to many types of farm animals, even when those animals are perfectly healthy. Because some of these same drugs are used for humans, such overuse quickly breeds resistant bacteria that can harm people.

Gut Flora

In recent decades, scientists have gradually become more aware of the importance of the microorganisms that live on and in our bodies, especially the ones that live in our intestinal tracts. A catchy name for these organisms is *gut flora*, a phrase that makes me picture a flower garden in my intestines. However, *gut flora* is certainly more concise and easier to remember than *human gastrointestinal microbiota*, so I think I'll stick with the shorter phrase.

More than 1,000 species of microorganisms live in human intestines around the world (along with a few species in people's stomachs). These are mostly bacteria but also include fungi, protists, and viruses. The exact mix of species varies from one person to another, with an average of 160 species in each person's gut. You could say that we each have our own customized gut flora—our own unique intestinal garden. The various species are present in differing quantities, with 99 percent of the bacteria in your gut consisting of just 30 or 40 species. Yet it seems that only 5 species are universal, found in the guts of all humans. One of those 5 is the well-known *Escherichia coli*, often called *E. coli*. The other 4 universal species are three types of *Bacteroides* and one of *Enterococcus*.

One benefit of your gut flora is that it helps to digest your food. Microorganisms in the intestines break down some of the hard-to-digest materials that would not otherwise be utilized. Another benefit is that by completely colonizing the available surfaces inside your intestines, your gut flora prevents harmful organisms from growing there. Still another benefit is that the gut flora appears to assist in the proper functioning of your immune system.

When you take an antibiotic, the goal is to kill off a particular species of bacteria that is suspected of causing a human disease. However, most antibiotics kill a wide range of bacterial species, which means that some of the helpful bacteria in your gut are also killed off, depleting your gut flora. Until a balanced gut flora can reestablish itself, your digestion may be less efficient, causing some intestinal distress. During this same period of time, you may also be at greater risk that some type of harmful microorganism, perhaps a fungus, might proliferate in your intestines.

Some people take probiotics after taking an antibiotic in an effort to restore the depleted gut flora. The word *probiotic* refers to a dose of live cultured bacteria. The word *cultured* means that the microorganisms were grown intentionally, typically in a vat. Some evidence indeed shows that taking probiotics is helpful in this situation, reducing the side effects of taking an antibiotic. In particular, seeding your intestines with harmless bacteria may be a good way of preventing harmful microorganisms from occupying the same sites. However, taking probiotics cannot restore your gut flora to its previous state, for the simple reason that we don't know how to culture most of the thirty or forty species of bacteria that dominate your gut flora. In fact, it's unlikely that anyone has even tried to identify which thirty or forty species constitute the bulk of your own personal gut flora. Instead, commercial probiotics are based on a tiny handful of bacterial species that people have found easy to culture.

Humans have a long history of culturing milk products such as cheese, yogurt, and sour cream, which means we're very good at growing the various microorganisms that enjoy milk. Thus, the probiotics available for sale are often the same bacteria used to turn milk into yogurt—including several species of *Lactobacillus*. In the future, as we learn more about which species of gut bacteria are valuable for maintaining good health, and as we learn how to culture those bacteria, probiotics might improve and diversify. The day might come when doctors can diagnose exactly what your gut flora is missing and prescribe the correct species of bacteria to restore balance to your gut. But we're not there yet.

As I finished my nightly ration of yogurt, with its live *Lactobacillus* and other species of bacteria, I had to end my evening's daydreams about microbes. But before leaving the kitchen, I decided to take an inventory of the cultured food products in the house—products that were made with the assistance of microorganisms.

In addition to yogurt, I soon found several other cultured dairy products in the refrigerator, including sour cream and three varieties of cheese. To make cheese, fungi are used instead of bacteria, and each type of cheese needs a specific type of fungus to develop the correct flavor. In the pantry I found a bottle of cooking wine. Wine is made by growing yeast in crushed grapes, and beer by growing yeast in sprouted grain. Yeast is a very interesting little microbe, considered to be a type of fungus. I then found a bottle of vinegar. Vinegar is made with the assistance of bacteria. I found a bottle of soy sauce, which is made by growing fungus on crushed soybeans. I found a loaf of bread, which is fluffy due to the action of yeast that makes the dough rise before the loaf is baked. I quickly created a mental list of other cultured food products not currently in my kitchen, including sauerkraut, kimchi, kombucha, miso, kefir, and tempeh, all of which involve culturing either bacteria or fungi.

This left me thinking. How nice it is that these microorganisms provide me with such a variety of flavors and textures in my foods! How nice it is that the microorganisms in my gut help to keep me healthy! Although I certainly intend to continue washing my hands on a regular basis, I don't need to think of all microbes as my enemy. Instead, I can think more broadly, beyond the simplistic concept of killing germs. The world of microscopic life consists of both friends and enemies, along with a great number of innocent bystanders that are completely harmless to me. With that thought resonating in my head, I double-checked the yogurt to verify that I had left it glassy smooth and headed back upstairs.

7

Twenty-Four Hours a Day

Throughout my childhood and well into adulthood, I tended to eat my meals in a highly systematic manner. This was not so apparent when I ate from a bowl (breakfast cereal or soup) or when I ate a sandwich, but it was extremely obvious whenever I ate dinner served on a plate. Of all my curious quirks, this was the one that generated the largest number of remarks from other people. And yet for me, it was all completely logical.

The remarks came because people noticed that I ate one item at a time, completely finishing each thing on the plate before moving to the next. A few people, more observant than the rest, noticed that I always ate my food in a clockwise direction. But there was more to it than that. To me, the food arrayed on the plate was like a pie chart, consisting of distinct zones divided by radial lines. Each type of food occupied one of those zones. To begin eating, I would carefully select one of the radii that separated two zones and then eat clockwise from there. I perceived an active eating radius sweeping around the plate, just like a second hand sweeping around a clock, and every forkful I took came from that radius. With my left hand, I rotated the plate as I ate so that the eating radius remained directly in front of me. As far as I was concerned, this was the only possible way to eat. I could not imagine myself eating in any other manner.

Of course, sometimes the food was served in a pattern that didn't precisely conform to a pie chart. This was usually easy to fix. Before taking my first bite, I would nudge the food with my fork until it was all properly arranged. However, it was a big problem when a food item completely covered the center point of the plate. This happened only on rare occasions, but it was quite frustrating when it did occur, and very hard to fix.

One of the comforting things about my style of eating was that it made me feel more connected to the solar system. The rotating plate, the rotating hands of the clock on the wall, the rotating Earth, and the revolution of the earth around the sun all felt connected. The clock on the wall—with the continuous sweep of its second, minute, and hour hands—precisely marked each period of twenty-four hours. I knew that light from the sun illuminates half of the world at any given moment, and I pictured this zone of sunshine sweeping around the earth in exactly twenty-four hours in harmony with the clock. Furthermore, I pictured the earth sweeping around the sun in a circular orbit, like a giant echo of the planet's rotation on its axis. As I completed each sweep around my dinner plate, I could imagine the earth completing another year of its orbit.

Except, as I learned much later, none of this is quite true. The clockwork of the solar system doesn't actually match that of the clock on the wall. The apparent precision with which we measure a twenty-four-hour day is an illusion created by humans for our own convenience. In reality, everything is a little bit out of kilter.

Our Concept of the Twenty-Four-Hour Day

We all assume, because this is what we're taught, that a day represents one rotation of the earth on its axis and is exactly twenty-four hours long. Our system of telling time is based on this idea; in fact, it's a core principle of our modern technological society. The high-tech machinery that runs our world relies on the precise measurement of time and the exact determination of the current time at any given moment. Our personal computers and phones measure time to the millisecond (1/1000 of a second), and the official atomic clocks (used to establish the standard time around the world) are precise to the billionth of a second. As a result, the concept of a twenty-four-hour day has become fully ingrained in our culture. For example, we commonly use the phrase "twenty-four hours a day" to indicate that a particular process continues nonstop. The upshot is that we accept without question the idea that one day equals exactly twenty-four hours (down to the millisecond), with every day having precisely this same length.

Thus, it might surprise you to learn that one spin of the earth on its axis does *not* take precisely twenty-four hours—and that the discrepancy amounts to several minutes every day. The mismatch arises because our traditional concept of a day is actually defined by the cycle of sunlight and darkness, and not by one rotation of the earth. This cycle is affected not only by the rotation of the earth on its axis but also by the revolution of the earth around the sun. It might also surprise you to learn that the time required for a cycle of daylight is not consistent but varies by nearly a full minute during the course of a year, a result of the earth's tilted axis and its slightly lopsided orbit around the sun. We only pretend that all days are the same length, by averaging the length of every day in the year and then defining this average as a standard day of exactly twenty-four hours.

This is not a bad thing. In fact, it has been quite helpful to define our system of time in this manner. But once you understand why this system doesn't quite match up with the real world, you can begin to make sense of several interesting phenomena. For example, you would think that the shortest day of the year—winter solstice, the first day of winter—would have the latest sunrise and the earliest sunset of the year, but this is not the case. (On the other hand, if instead of clock time we all used sun time, as measured by a local sundial, it would indeed be true.)

If our definition of a day was actually based on one complete rotation of the earth on its axis, a 360-degree spin, a day would be nearly four minutes shorter than our standard twenty-four-hour day. This is because the earth must spin on its axis *more* than 360 degrees between one dawn and the next as it orbits around the sun. During a single day, the earth travels about 1/365 of the way around the sun, because it takes about 365 days (one year) to go all the way around. This daily progress in the earth's orbit is almost exactly 1 degree of travel (defined as 1/360 of a circle). It follows that the earth has to spin on its axis an extra degree each day in order for any particular spot on the earth to line up with the sun again. Therefore one complete cycle of sunlight and darkness—one day—represents a rotation of roughly 361 degrees.

Although a year consists of 365¼ days, Earth actually spins on its axis 366¼ times during a year. From the standpoint of sunrises and sunsets, one complete spin is negated each year by the journey around the sun. The result is that the earth must spin one extra time each year compared to the annual number of sunrises and sunsets. If the earth did not spin on its axis at all, we would still have night and

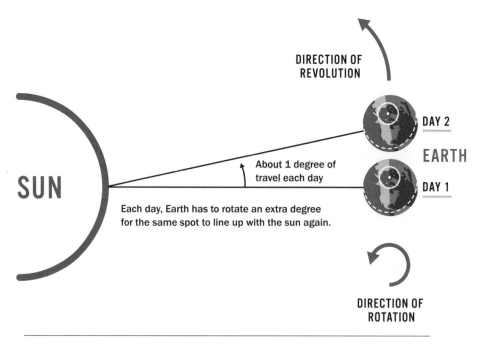

DIRECTION OF
REVOLUTION

DAY 2

EARTH

About 1 degree of
travel each day

SUN

DAY 1

Each day, Earth has to rotate an extra degree
for the same spot to line up with the sun again.

DIRECTION OF
ROTATION

Why a twenty-four-hour day is slightly more than a full spin of the earth day on this planet; it's just that a day would be the same length as a year, because that's how long it would take the sun to rise and set one time. Furthermore, the sun would rise in the west and set in the east, the opposite of what we see now.

A Stellar Day

Another way of expressing this idea is that the length of a day depends on our frame of reference. For reasons both traditional and logical, we normally use the sun as our reference when defining a day. But if you wanted to define a day as one complete spin of Earth on its axis, you could use the stars as your reference. By observing the locations of the stars in the sky, you can determine when the earth has completed exactly one rotation. We call a day defined in this manner a stellar day or a sidereal day. (The two terms have slightly different definitions but nearly identical results.)

In a stellar day, any specific star rises at approximately the same time every day. More precisely, this star reaches its highest point in the sky at the same time each stellar day. This is because at the beginning and end of one complete rotation of the earth, any given point on Earth now faces the same direction in space. (In other words, any point on Earth now faces the same direction with reference to the other stars in our galaxy.) But because a stellar day is about four minutes shorter than a traditional solar day, any given star rises about four minutes earlier each solar day. For example, if you enjoy spotting the constellation Orion in the autumn and winter, you probably notice that it rises slightly earlier each night. If you go out at precisely ten o'clock each night to note the position of Orion, you'll see that every night the constellation is a little bit higher in the eastern sky than the night before.

Unlike a solar day, whose true length varies throughout the year, a stellar day is always twenty-three hours, fifty-six minutes, and four seconds. However, if you want to be precise to the millisecond, you have to consider several kinds of wobbles that affect the direction of Earth's axis. (The largest of these wobbles is a cycle of twenty-six thousand years called an axial precession.) You also have to consider the very gradual slowing of the earth's spin. (One complete rotation takes about 1.7 milliseconds longer than it did a century ago.)

A Solar Day

In contrast to a stellar day, a solar day—one complete cycle of sunlight and darkness—is much more variable in length. However, the amount of variability depends in part on when you consider the day to begin. For example, you can consider a new day as beginning at dawn, when the sun rises, or you can consider the day as ending when the sun sets—in which case the next day begins at sunset. (Several prominent religions still use the latter system.) A third choice is to say that the new day begins at the exact middle of the night, halfway between sunset and sunrise. (This moment can be called true midnight.) In any of these systems, the length of a day varies throughout the year, but the average length is twenty-four hours. The midnight system exhibits considerably less variability in day length than either the dawn or the dusk systems.

In our current system of standard time, a new day does indeed begin at midnight—except that it's not true midnight. At any given location on any given day, the difference between true midnight and midnight according to standard time can be significant. In the United States, the difference can be as great as an hour; and during the months of daylight saving time, the difference can even reach two hours. In far western China, the difference between standard time and true time is three hours, a result of stuffing the entire country into a single time zone.

Although you can consider the length of a true solar day to be the amount of time from one true midnight until the next true midnight, a reasonable alternative is to consider it as the amount of time from one true noon until the next true noon. At any given location, true noon occurs halfway between sunrise and sunset (assuming a relatively flat horizon, without mountains). More precisely, it is the moment when the sun reaches its highest point in the sky for that day. In the northern temperate regions of the world, the sun is due south at true noon. In the southern temperate regions of the world, the sun is due north at true noon. In the tropics—that is, any place in the world that's south of the Tropic of Cancer but north of the Tropic of Capricorn (more on this later)—the sun is either due north or due south at true noon, depending on which day of the year it is. Furthermore, twice a year the sun is directly overhead at true noon at any given location in the tropics. The exact dates of this phenomenon vary according to the latitude of the location.

This point was driven home to me the year I lived in a house very near the equator—about 5 degrees north latitude. In early January, soon after moving into that house, I planted some shade-loving plants on the north side of the house, where the shadow of the house protected my plants from the intense midday sun. The plants thrived in that location—for a few months. But six months later the north side of the house was the sunny side, while the south side was the shady side. My shade-loving plants all gave up the ghost.

Two independent factors cause the length of a solar day to vary from the twenty-four-hour average. The first factor is Earth's orbit around the sun, which varies in distance and speed. The orbit is not a perfect circle; in fact, the distance between the earth and the sun changes during the year by about 3 percent (roughly 3 million miles). The speed at which the earth travels around the sun also varies by about 3 percent during the year and is fastest when the earth is closest to the sun. The changing distance and speed both affect the length of a true solar day, because they affect how many degrees the earth must spin between two consecutive instances of true noon at any given location. This factor adds about ten seconds to the length of a solar day at the beginning of January, when the earth is closest to the sun, and it subtracts about ten seconds from each solar day at the beginning of July, when the earth is farthest from the sun.

The second factor that affects the length of a solar day is the tilt of the earth's axis, for reasons I won't try to explain. (In this case, a need for brevity will trump my desire for thoroughness.) The result adds about twenty seconds to each solar day at the summer and winter solstices (in June and December) and subtracts about twenty seconds from each solar day at the equinoxes (in March and September). During parts of the year these two factors (the lopsided orbit and the axial tilt) reinforce each other, and at other times these two factors partially cancel each other out.

The net result is that the length of a solar day reaches a maximum of twenty-four hours plus thirty seconds in late December, with a second smaller peak in June. In mid-September, the length of a solar day reaches a minimum of twenty-four hours minus twenty-one seconds, with a second minimum in late March. Thus, the length of a solar day varies by nearly a minute during the year.

Although this effect is relatively small when you consider only a single day, it is quite noticeable when accumulated over several months. Imagine you have two clocks. The first clock, which is simply a very accurate sundial, shows true local time, including true noon. The second clock is electric, but instead of being set to standard time

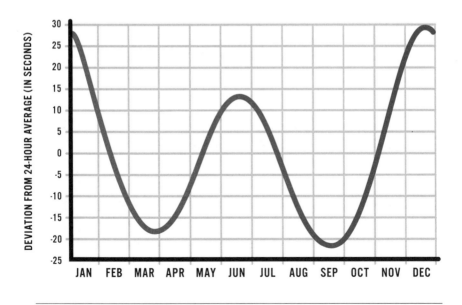

Annual variation in the length of a solar day

according to the local time zone, it is set to local mean time. (The word *mean* in this case means average and is not a reference to the cruelty of time.) In other words, the electric clock assumes that all days are exactly twenty-four hours long but that the *average* time of noon should match up with the local true noon as indicated by the sundial. Four times a year, the two clocks will agree as to when it is noon. But in early February, the electric clock indicates noon a full fourteen minutes before the sundial does. In early November, the electric clock indicates noon sixteen minutes later than the sundial does. This is a rather stark difference!

This mismatch is why the latest sunrise and the earliest sunset of the year do not coincide with the shortest period of daylight during the year—the winter solstice, which occurs on or near December 21 in the Northern Hemisphere. (In the Southern Hemisphere, this same date has the longest period of daylight and is therefore the summer solstice.) If we actually used true sun time (in contrast to either local mean time or standard time), the latest sunrise and earliest sunset of the year would indeed occur at the winter solstice, on the first day of winter.

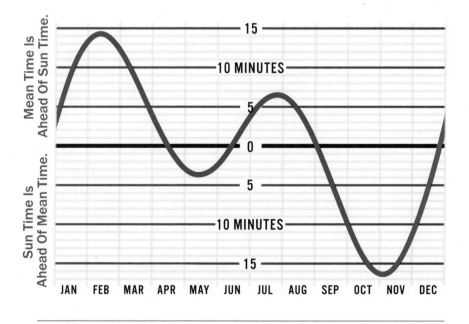

Deviation between sun time and mean time

People often describe the winter solstice as "the shortest day of the year," and if you measure a day as lasting from sunrise to sunset, this description is absolutely true—at least for the temperate zones of the world. Thus the shortest period of daylight in the United States is in December, while the shortest period of daylight in Australia is in June. But if you measure the length of a day from one true noon to the next, the shortest day of the year is in September, regardless of where you live. Therefore the phrase "shortest day of the year" is rather ambiguous.

The Solstices, the Equinoxes, and the Analemma

Of course, the only reason we even have a summer and winter solstice is that the earth's axis is tilted by 23.5 degrees compared to the axis of our orbit around the sun. But what exactly do we mean by *axis*? Our planet rotates around an invisible line that runs through its center, from the North Pole to the South Pole. In fact, this is how we define the poles, as the only two points on the surface of the earth that lie on the axis of spin. This axis always points in the same approximate direction relative to the stars. In the Northern Hemisphere, the axis points almost directly to Polaris, which we call the North Star. In the northern night sky, all the other stars slowly rotate around Polaris, a phenomenon clearly visible in time-lapse photos.

In late December, the North Pole is tilted 23.5 degrees away from sun, while the South Pole is tilted 23.5 degrees toward the sun. At this time of year, the sun never rises above the horizon at the North Pole, and it never sets below the horizon at the South Pole. In late June, at the other solstice, the reverse occurs. But at the two equinoxes in March and September, the poles are not tilted either toward the sun or away from it. However, if you view the earth from the direction of the sun, both poles are tilted sideways 23.5 degrees. Furthermore, the equator is tilted 23.5 degrees compared to the path of Earth's orbit.

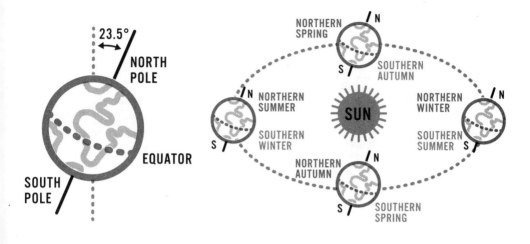

The tilt of the earth's axis

At any given moment, the sun is directly overhead at a single point on the earth. If our planet's axis did not tilt, the sun would always be directly overhead at some point on the equator. Instead, the spot where the sun is directly overhead gets as far north as 23.5 degrees north latitude (at the June solstice) and as far south as 23.5 degrees south latitude (at the December solstice); the sun is directly overhead at the equator only on the two equinoxes. Notice that the latitudes at the two solstices exactly match the tilt of the earth's axis. We call these lines of latitude the Tropic of Cancer (in the north) and the Tropic of Capricorn (in the south), and the zone in between them the tropics. As hinted at earlier, unless you live in the tropics, the sun is never directly overhead where you live. In the continental United States and Europe, the sun is always due south at true noon, all year long.

Imagine that one fine day, at a moment when the sun is directly overhead at a point in the Pacific Ocean, you put a dot on the globe indicating that point. Exactly twenty-four hours later, you place another dot on the globe indicating the current position of the sun. You continue this process at twenty-four-hour intervals for a full year. What will the final result be?

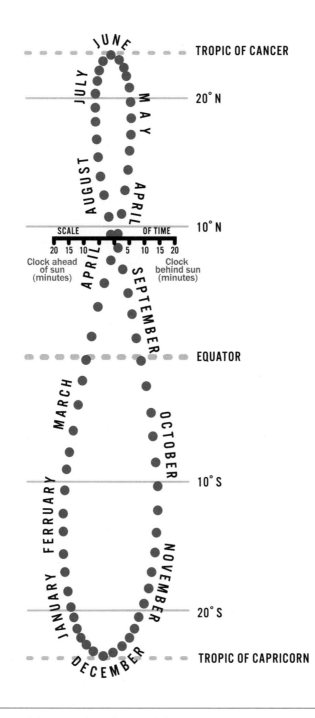

Analemma of the sun, projected onto a globe

First of all, you know that the dots will stretch between the Tropic of Cancer and the Tropic of Capricorn. Second, you know that if all solar days were exactly twenty-four hours long, the dots would form a straight north-south line on the globe. But because solar days are variable in length, sometimes a bit longer than twenty-four hours and sometimes a bit shorter, a different shape will appear instead.

In fact, your dots will form a lopsided figure eight on the globe. Many globes have such a figure printed on them. This mysterious shape, called an analemma, used to puzzle me when I was a kid. I wondered what it meant, and why it was important enough to appear on the globe. Furthermore, I was quite bothered by the asymmetry, with one lobe of the figure bigger than the other. Many decades later I finally understood, and now you do too. The asymmetry, which is the hardest part to explain, is due to the asymmetric combined effects of the two factors (elliptical orbit and axial tilt) that cause a solar day to deviate from twenty-four hours.

Time Zones

The use of true sun time was abandoned when mechanical clocks allowed us—or perhaps forced us—to adopt a standard length of twenty-four hours per day, regardless of the actual position of the sun. Thus, sun time was replaced by local mean time three centuries ago. But of course, we don't use local mean time anymore either. If we did, whenever you traveled a few miles east or west, you would have to adjust your watch. Ever since the 1880s, we have relied instead on a standard time system based on time zones. The idea is to divide the world into twenty-four north-south stripes or zones. Within each zone, everyone uses the same time. As you go from one zone to another, you typically adjust your clocks by exactly one hour. Each time zone has an average width of 15 degrees of longitude (the result of dividing 360 degrees into twenty-four equal pieces). This translates to a width of about 1,000 miles at the equator, but the time zones get progressively

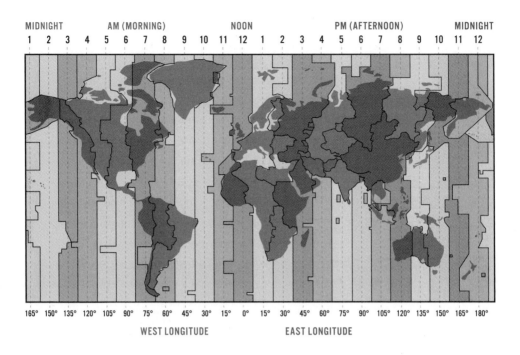

The time zones of the world

narrower as you travel toward either of Earth's poles—and approach a width of zero at the poles themselves, where all the time zones converge. However, you wouldn't know this from looking at some of our world maps, which show lines of longitude as parallel lines.

Within each time zone, standard time is based on the local mean time at a specific longitude within the zone. For example, in the eastern time zone in North America, standard time is based on 75 degrees west longitude, the theoretical center line for this time zone. In any city or town with this same approximate longitude, such as Philadelphia, local mean time matches standard time nearly perfectly. But in a city or town with a longitude of around 80 degrees west, such as Charleston, South Carolina, standard time and local mean time differ by around twenty minutes. In Indianapolis, located at about 86 degrees west longitude, the difference between standard time and local mean time is approximately forty-five minutes. During the summer, when daylight saving time (DST) is in effect, the deviation increases by another hour. The result is that in the summer, standard

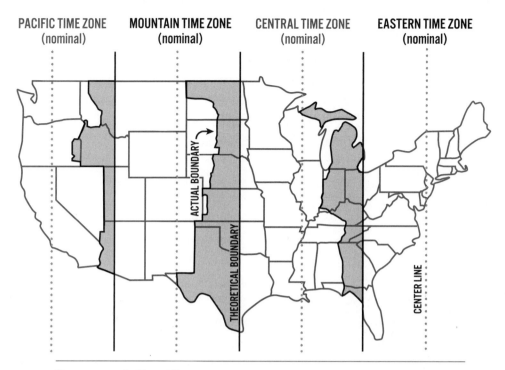

PACIFIC TIME ZONE (nominal)   MOUNTAIN TIME ZONE (nominal)   CENTRAL TIME ZONE (nominal)   EASTERN TIME ZONE (nominal)

ACTUAL BOUNDARY   THEORETICAL BOUNDARY   CENTER LINE

Time zones in the United States

time and local mean time differ by nearly two hours in Indianapolis. (For locations that are east of the center line, such as Boston, DST causes a flip from a small negative deviation to a somewhat larger positive deviation.)

Indianapolis used to be in the central time zone, along with the entire state of Indiana, but now most of the state is in the eastern time zone. If Indianapolis still used central time, standard time and local mean time would differ by only fifteen minutes instead of forty-five minutes. The problem is that on average the sun would already be halfway across the sky by 11:45 a.m. In early November, because of the difference between mean time and actual sun time, the sun would be halfway across the sky at 11:30 a.m., which is awfully early for half the day to be gone! Likewise, dawn and sunset would both occur a half hour earlier than expected. It seems that most people would prefer to have the sun halfway across the sky at 12:30 p.m. rather than 11:30 a.m. The solution was to put most of Indiana into the eastern time zone. A similar issue affects any other place located to the east of the center line of the corresponding time zone. The result is that politicians are constantly tempted to shift the time zone boundaries westward. A majority of places in the United States (but certainly not all) are now located to the west of the center line for the corresponding time zone. In the map on the previous page, all the areas shaded gray have been shifted from their theoretical time zones into the neighboring zone to the east.

Seasons of the Year

Why is it hotter in the summer than in the winter? A surprisingly large number of people say it's because we're closer to the sun in the summer, which is completely wrong. Many of these same people recognize that the correct answer involves the tilt of the earth's axis—ten bonus points for that! To connect the two concepts, they explain that the earth's tilt puts part of the earth farther from the sun than the rest of the earth, and this greater distance causes winter. Nice try, but that irritating sound you just heard is the wrong-answer buzzer.

Of course, the wrongest of the wrong answers is to say that the entire earth is closer to the sun in the summer. This idea is self-contradictory because when it's summer in the Northern Hemisphere, it's winter in the Southern Hemisphere, and vice versa. As a kid, my smarty-pants response to this "closer-to-the-sun" idea was to say that the earth is always the same distance from the sun—93 million miles—and therefore the earth could not possibly be closer to the sun during the summer.

It turns out that I was wrong too. Oh, well! I eventually learned that the earth's orbit deviates significantly from a perfect circle. As a result, the distance between the earth and the sun varies by 3 million miles during the course of each year. That's enough to make a noticeable difference in temperatures. (In contrast, the tilt of the earth's axis puts the wintery parts of the earth only about 2,000 miles farther from the sun, which makes no difference whatsoever.) The big catch in all of this is that the earth's closest approach to the sun (called the perihelion) occurs in early January, during the heart of winter in the Northern Hemisphere.

Huh? This means that those of us living in the Northern Hemisphere are 3 million miles closer to the sun *during winter* than we are during summer! Where is the logic in that?

The explanation is that our cycle of seasons is indeed due to the tilt of the earth's axis, which has a big impact on the intensity of sunlight striking Earth—but not because the tilt affects our distance from the sun. The intensity of sunlight is strongly affected by how high the sun is above the horizon. When the sun is low in the sky, the rays strike the earth at an oblique angle, spreading the energy over a large area and thereby decreasing the intensity. Thus, sunlight is much more intense at noon than it is in the early morning or late afternoon. Likewise, sunlight is more intense in the summer, when the sun rises higher in the sky, than it is in winter. There are also more hours of daylight in the summer, which results in a longer daily period of heating and a shorter nightly period of cooling. For most of the world, the total amount of solar energy striking the earth is much greater in summer than in winter.

In the tropics, the height of the sun at noon does not change much during the year, nor does the length of the daylight period. Therefore the tropics don't experience much variation in average temperatures during the year.

The varying distance between the earth and the sun does have an effect on our seasons, but perhaps not in the way you expected. The sunlight falling on Earth is 7 percent more intense in early January than in early July. In the Northern Hemisphere, this variation lessens the temperature difference between summer and winter; in the Southern Hemisphere, it exaggerates the temperature difference between summer and winter.

Adding Up the Sunshine in a Day

As mentioned earlier, two important factors affect the total amount of solar energy that falls on any given spot on any given day: the angle of the sun in the sky and the length of time the sun is above the horizon. (The degree of cloudiness also plays a big role, but for the purpose of this discussion let's assume a beautiful sunny day.) As you go farther from the equator at the height of summer, these two factors pull in opposite directions, with the sun getting lower in the sky but the days growing longer. For a period of several weeks centered on the summer solstice, these two factors essentially cancel each other out—at least in the temperate zones of the world. During this time period, the total amount of daily sunshine is about the same across a wide range of latitudes. For example, in the early summer (June and the first half of July), a plot of land in North Dakota or Manitoba receives about the same amount of daily solar energy as a similar plot of land in Texas.

If you need a term to describe this concept, a good choice is *daily insolation*, although the latter word could easily be confused with *insulation*, which is a very different concept. (An alternative term is *daily solar irradiation*, which seems rather awkward and slightly scary.) While the daily insolation is fairly consistent from place to place in early summer, the same is certainly not true in early winter. At this time of year, places farther from the equator have shorter days *and* the sun is lower in the sky. Thus, in winter your latitude has a huge impact on the amount of daily insolation. The result is that a summer day can be hot at nearly any latitude, but the temperature of a typical winter day varies considerably with the latitude.

This effect is much stronger in locations that have a continental climate, as opposed to a marine climate. Oceans and other large bodies of water serve as heat sinks, moderating the temperature swings in nearby land areas. In other words, locations that are far from the equator *and* far from any ocean will experience the greatest difference between summer and winter temperatures. Siberia, located on the giant continent of Asia, is the prime example of such extremes, but the northern plains of North America (such as North Dakota and southern Saskatchewan) also provide a good example. And finally, places with low humidity and little cloudiness (that is, most deserts) experience greater daily and seasonal temperature swings than cloudy, humid places, because they lack the insulating effect that clouds and humidity provide. (Did you notice that? We've just brought in insulation as well as insolation!)

The Clockwork That Isn't Quite So Clocklike

To review, let's return to our original question: How long is a day? If we define a day based on an actual, daily, observable physical phenomenon, we have two leading possibilities for the answer. If we define a day as one complete spin of Earth on its axis (a stellar day), a day is about twenty-four hours minus four minutes long. If we define a day as the time between true noon one day and true noon the next day (a solar day), the length of a day varies throughout the year; at one extreme it's thirty seconds longer than the twenty-four-hour average, and at the other extreme it's twenty-one seconds shorter. But if we take the average length of all the solar days in a year, the result is exactly twenty-four hours, which is how we arrived at our standard day. The upshot is that the standard twenty-four-hour day is not something found in nature but a human invention that only roughly corresponds to the real days— solar or stellar—we actually experience on Earth.

Tonight I am contemplating these issues as I eat my dinner. I used to feel bothered that the clockwork of the solar system is not quite as neat and regular as I had once assumed. Days are not actually all the same length; the orbit of the earth around the sun is not actually a perfect circle; and the speed of the earth around the sun is not perfectly consistent. With this in mind, I now try to be less systematic when I eat. (Okay, I'll admit that social pressure is also a factor in my changed eating habits.) That said, I still sometimes see those radii dividing my plate into zones, and I still sometimes feel the temptation to rotate the plate as I eat. I won't publicly admit it, but when I eat alone, it's possible that I occasionally give in to those temptations, eating one item at a time in a delightfully rigorous sweep around the plate. Yes, a comforting thought indeed!

8

The Blueprint of Life

My family and friends are well aware of my peculiar habits, which reveal a bit of an obsessive-compulsive streak in my personality. Some of my friends jokingly refer to this behavior as OCD (obsessive-compulsive disorder), but this labeling conflates a slightly odd but normal personality type (such as mine) with a psychological disorder. (Admittedly, there is a continuum between the normal and the abnormal, with no sharp dividing line between them.) I also see some irony in this use of the word *disorder*, because I have a distinct dislike of disorder. Instead, my habits are all about maintaining a state of *order*. Therefore, I would prefer to describe my behavioral traits as OCO—obsessive-compulsive order. And of course, it is this OCO that causes me to be such a stickler for getting the details right, especially when it comes to how writers in the popular media cover science.

As a prime example of this phenomenon, it distresses me whenever I see DNA described as "the blueprint of life." This is a terrible metaphor because it is highly misleading, and yet the underlying idea is not 100 percent wrong. The metaphor correctly implies that the information contained in DNA is responsible for every form of life on this planet, including everything now living and everything that has ever gone extinct. But beyond that, this metaphor leads our thinking down the wrong path, which makes it a poor choice for journalists to use.

If this is not a good metaphor, is there a better one? Some people say DNA is like a computer program, while others say DNA is like a recipe or a list of ingredients. Let's examine these competing metaphors to consider the strengths and weaknesses of each, and to see whether any of them can help us to understand what DNA is all about.

Blueprint, Computer Program, or List of Ingredients?

The term *DNA* (short for *deoxyribonucleic acid*) appears often in the media. Because of these frequent encounters, we all have a general idea of what the term means. When we think of DNA, we think of genes, and therefore we correctly associate DNA with genetic inheritance. In particular, we think of the human genome, which includes the complete set of genes in a human being, distributed across twenty-three pairs of chromosomes. Yet our common understanding of DNA tends to be highly limited and somewhat inaccurate. Consider the following three statements. Which one of these assertions would you judge to be the most accurate?

- **ASSERTION 1:** The human genome is much like a set of blueprints. Our cells use these blueprints as a guide for constructing the human body.
- **ASSERTION 2:** The human genome operates like a computer program, with coded instructions for building and maintaining a human body.
- **ASSERTION 3:** The human genome is primarily a set of lists that specify the sequence of ingredients for assembling proteins. Our DNA does not actually contain any plans or instructions for building a human body.

There are good reasons to quibble with all three of these assertions, but the most accurate of the three is assertion 3. The human genome is not at all like a blueprint, as we shall see. The comparison with a computer program is somewhat more enlightening but still highly misleading. But it is indeed true that the human genome is primarily a collection of ingredient lists for assembling proteins. DNA is actually a bit more than that; it also includes templates for RNA molecules that help regulate the manufacture of these proteins. Thus, DNA is all about making proteins, without any specific plans or instructions as to how these proteins are supposed to yield a human body—or an elephant, or an oak tree.

Assertion 3, although mostly accurate, can still mislead our thinking, primarily because the word *proteins* can send our thoughts in the wrong direction. The missing concept is that proteins control most of the development, structure, and maintenance of the human body. Our DNA, due to its central role in the creation of proteins, *indirectly* controls these processes. To make sense of this idea, we first need to take a closer look at the wide variety of proteins in your body, along with the crucial roles these proteins play. Then we can return to the question of how DNA actually works.

The Essential Roles of Proteins in Your Body

Most of us think of protein as the material that muscles are made of. In the popular imagination, this is the sole reason proteins in your diet help to build and maintain a strong body. It is indeed true that your muscles are primarily made of protein—not just the skeletal muscles that allow you to move around but also the muscles associated with internal organs, such as the heart. However, the human body actually contains many different kinds of proteins, performing several distinct roles:

- **ENZYMES** drive most of the chemical reactions that occur inside your body and your cells. Thousands of such reactions take place, with an incredibly wide range of results. Enzymes produce most of the chemical compounds needed by your body, and enzymes played a huge role in the development of your body from a single-celled zygote to a fully formed human being. For example, the cell membranes in your body are primarily constructed from lipid molecules, a multistep process driven by several different enzymes.
- **SIGNALING AND REGULATORY PROTEINS** carry signals between different parts of the body, coordinating biological processes that involve multiple cells. For example, the hormone insulin is a protein that tells the cells in your body when to absorb glucose from the blood, thereby regulating your blood sugar level.
- **TRANSPORT AND STORAGE PROTEINS** move small but vital molecules from one place to another, either within cells, across cell membranes, or between cells. These proteins also help store some of these materials. A great example of a transport protein is hemoglobin, found in red blood cells, which carries molecules of oxygen from your lungs to all parts of your body.
- **DEFENSE PROTEINS**, such as antibodies, are mostly found in your blood, where they bind to foreign particles and molecules, including viruses and bacteria, thereby disabling them.
- **STRUCTURAL AND CONTRACTILE PROTEINS** serve roles such as forming the connective framework of your muscles, bones, tendons, skin, and cartilage; providing physical support for cells; and allowing parts of the body to move (due to the ability of muscle cells to contract).

In short, proteins control most of the processes that occur in the human body. Proteins keep your body running and (rather amazingly) mediate most of the human development process. But it is your DNA that tells your body which proteins to build.

DNA as a Template for Proteins

You have approximately twenty thousand genes in your DNA, and two copies of most of these genes. For some of your genes, the two copies are identical, but in many cases the two copies have subtle differences. But what exactly is a *gene*? The term was invented before the discovery of DNA, but now we think of a gene as a coded DNA template for assembling a specific protein. For the most part, each gene is unique, which implies that your DNA contains the instructions for building as many as twenty thousand different proteins. However, the picture is actually more complicated than this, because some proteins are built by using only selected parts of a gene and leaving out other parts. As a result, a single gene can often serve as a template for several variant proteins. Another complication is that some proteins are assembled by linking together two or more smaller proteins, each separately coded in the DNA. Therefore, the number of distinct proteins your body can build is probably at least ninety thousand.

So what exactly is a *protein*? A protein is a long chain of amino acids strung together in a linear sequence. Amino acids are rather small molecules, each containing between ten and twenty-seven atoms. Twenty distinct amino acids are typically used for assembling proteins (although under certain circumstances two other amino acids can also be used). These 20 amino acids have names such as *leucine*, *glutamine*, and *tryptophan*. Each protein consists of a specific sequence of these amino acids. A typical protein is a chain of 250 to 500 amino acids, but some proteins can be shorter or much longer. Most proteins fold up into a compact shape as soon as they're created, and the specific role the protein plays is often connected to the shape of the folded molecule. Thus, the final shape of a protein is more likely to resemble a crumpled ball of paper than a chain or necklace.

Each gene in your DNA is actually a list of amino acids, like a shopping list or a list of ingredients, written in code. This list tells your body which amino acids to use when it builds a specific protein molecule. The list also indicates the exact sequence in which these amino acids should be strung together. In other words, the whole point of a gene is to contain *information*. This information is used by a living cell to build the appropriate proteins.

Inside each cell, embedded in the cytoplasm, is a collection of tiny protein-building machines called ribosomes. The odd thing is that nearly all of these ribosomes are identical. None of them are specialists. Any ribosome can build any kind of protein you ask it to build. You simply have to give the ribosome a coded message—copied from a gene in the DNA—that contains the recipe for the desired protein. The molecule that carries the coded message is called messenger RNA. The information in the message is just a long list of amino acids in the precise sequence for building the protein. If the required amino acids are available, the ribosome cranks out a brand new custom-built protein molecule according to the specifications contained in the message.

Note that DNA and RNA are very similar types of molecules. Your genetic inheritance is permanently stored in long DNA molecules. RNA molecules tend to be much shorter in both length and lifespan, serving as useful but temporary copies of certain stretches of your DNA.

The protein recipes are stored in your DNA using a distinct coding system. To make sense of the coding system, it helps to picture the DNA molecule. You have probably seen illustrations of the DNA double helix,

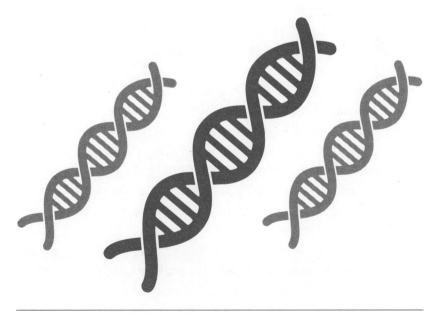

DNA molecules, shaped like spiral ladders

a spiraling ladder connected by evenly spaced rungs. The four possible kinds of rung are called A, C, G, and T (standing for *adenine, cytosine, guanine,* and *thymine*), and these rungs can potentially appear in any order. Any specific order of rungs has a specific meaning.

The core unit of information in a gene is a codon, equivalent to a three-letter word using this alphabet of just four letters (*A, C, G,* and *T*)—for example, GGA or TAC. Using the terminology of information science, we could say that each codon consists of exactly three bits of information, stored in three consecutive rungs, and that each bit can be any one of four distinct values. Using biological terminology, we would say that each rung contains one of four nucleotide molecules: adenine, cytosine, guanine, or thymine.

Now here is the fascinating thing about the DNA coding system. There are sixty-four different ways to string the four possible letters into a three-letter codon, and each of these sixty-four combinations has a specific meaning. The codon GGA means the next amino acid in the protein should be glycine. The codon CAC means the next amino acid should be histidine. Sixty-one of the sixty-four possible codons represent specific amino acids. But because there are only twenty amino acids, several different codons can have the same meaning. For example, CAA and CAG both indicate glutamine. The other three combinations (the ones that don't usually correspond to an amino acid) all mean "stop," which tells the ribosome that the protein is now complete.

Imagine you've discovered a gene for a protein that consists of exactly four hundred amino acids. How many rungs would you find in this stretch of DNA? Well, you would need four hundred codons for the amino acids, plus one for the stop code. Each codon requires 3 rungs, so the answer is 1,203 rungs in total. If you spelled out this gene as a set of letters, you would use 1,203 letters—one letter for each rung.

One of the most amazing things about this system of storing information is that all forms of life use the same DNA code—the same four letters mapping to the same twenty amino acids (with a few rare variations). That's what makes it possible to take a gene from one kind of creature and put it into the DNA of another kind of creature. If the gene in its new home actually gets used—that is, if copies of the gene are carried by messenger RNA to the cell's ribosomes—it will produce the exact same protein as always.

Not Like a Blueprint or a Computer Program

So why is it not so accurate to say DNA is like a blueprint for your body? A real blueprint—such as the plan for a house or an office building—is a detailed visual representation of *structure*. A blueprint shows all the structural components of the planned building, indicating the precise location and dimensions (and sometimes composition) of each part. In the blueprint, you can see the dimensions of every space, the location and size of every column and beam, the routes of the plumbing and the air ducts, and many other key details. Before construction even begins, this blueprint specifies the exact structure of the finished building, thereby allowing the project to proceed.

But DNA is nothing like that. DNA contains no depiction of the finished body. There is no drawing that shows the skeleton, no sketch indicating the locations of the internal organs, and no diagram illustrating the routes of the blood vessels and nerves. There is nothing that depicts what the finished face should look like. There is no set of specifications indicating how long the finished arm and leg bones should be, nor how fine or coarse the hair should be. DNA is simply a coded list of the amino acids for each protein the body can make, and somehow this information eventually yields the correct finished product: a human being, or an elephant, or an oak tree. DNA is ultimately responsible for most of the details but not in the manner the blueprint metaphor implies.

If DNA is not like a blueprint, is it like a computer program? After all, a computer program is completely different from a blueprint. Like DNA, a computer program contains no depiction of the finished result. Instead, a computer program precisely defines a *process*, a set of steps to execute. The process encoded by a computer program is typically called an algorithm. An algorithm is seldom a straightforward list of steps. In addition to being extraordinarily detailed, it is also full of loops and conditional branches. A loop is when the algorithm repeats a certain set of steps over and over. A conditional branch is when the algorithm tests a certain condition and then, based on the results of that test, either continues forward or jumps to a different point in the process.

To get a sense of what these loops and branches are like, imagine you have somehow acquired an industrial robot, similar to those on an automobile assembly line, and you want to program this robot to make drop cookies in your kitchen. You promptly set to work writing the program. When you get to the point where the robot deposits cookie dough by the spoonful onto a greased baking sheet (a highly repetitive process), you realize that the process consists of three nested loops—that is, a loop inside a loop inside a loop. The inner loop directs the robot to deposit dollops of cookie dough onto the baking sheet, one cookie at a time, until a row of dollops has been completed. The middle loop directs the robot to create one row of dollops at a time, until the baking sheet is filled. The outer loop directs the robot to fill one baking sheet at a time, until there is no more dough. Note that each of the three nested loops contains a distinct process to repeat, plus a test to determine when to stop looping.

Now let's compare a computer program to a blueprint. Suppose, as part of a woodworking project, you needed to create a square piece of plywood, 9 inches on each side. In a blueprint, you would see a nice drawing of a square, with a notation indicating that each edge is 9 inches long. In contrast, a computer algorithm might say:

1. Start with a large sheet of plywood.
2. With a circular saw, make a 9-inch straight cut in the plywood.
3. Turn the saw 90 degrees to the left.
4. Repeat the previous two steps three more times.

This algorithm does not depict the finished result; it doesn't even mention the idea of creating a square. But the specified steps do indeed result in a square piece of plywood, 9 inches on each side. Most of the processes in the human body are comparable in that the end product is not defined in advance; it simply emerges as a result of the process.

In one sense, the comparison of DNA to a computer program is enlightening. Countless biochemical processes are going on within the human body at all times, each contributing to the development and maintenance of that body. These processes certainly involve a lot of looping and branching, and therefore they could validly be compared to the algorithms encoded in a complex computer program. The problem with this analogy is that these processes are not actually encoded in your DNA. The language of DNA does not contain codes to indicate the start or the end of a loop, nor when a branch should occur, nor what test to perform when a branch point is reached, nor where in the instructions to jump based on the results of that test.

Worse yet for our computer program analogy, there are no DNA codes for *any* of the individual steps in any of the many thousands of biochemical processes. DNA is simply not capable of representing nor communicating this type of information. The codons in DNA provide no instructions whatsoever for any of these processes. The primary task of your genes is to provide lists of amino acids. Thus, DNA is not like a computer program at all.

More Like a Computer Data File or a Recipe

Although the analogy with a computer program is highly misleading, DNA can accurately be compared to a computer *data file*. Most computer programs are written to operate using data, which is stored separately from the program itself. If you use a word processing program, you save each document you write as a document file. But you can't open any of those document files without a computer program that knows how to read and interpret them. If you've ever downloaded a music file, what you downloaded was a data file. Besides the data file, you also need a program file, a computer program that serves as a music player, in order to open and play that music.

Each type of data file uses a different encoding system, but all of these data encoding systems are much simpler than a programming language, which is what we use to encode a computer program. As a result, the process of decoding and executing a computer program is far more elaborate than the process of decoding stored data. Nothing in DNA resembles a programming language, and nothing in living cells behaves like a computer's central processing unit—capable of reading, interpreting, and executing a computer program. DNA is very much like a data file but not at all like a computer program. Even though no computer program is involved, it is truly amazing that all of life is based on a data encoding system. Nothing else in living cells is comparable. DNA and RNA are the only molecules based on the encoding of pure information. The flexibility of this system allows for a nearly infinite variety of proteins to be built, all using the same ribosomes.

But if DNA is like a data file, doesn't that imply the existence of a computer program to read the data? No, not at all. Data can be read and interpreted by mechanical means that don't involve the use of computers, computer programs, or programming languages. In fact, mechanical devices that can read stored data are incorporated into record players, audiocassette players, videocassette recorders (VCRs), and DVD players. Within your cells, ribosomes use a biochemical system to interpret the data provided by messenger RNA (copied from DNA in the nucleus). Messenger RNA is very much like a movie film or an audio tape in that the data is read from end to end in sequence. The ribosome uses essentially mechanical means to detect one codon at a time and to use this information to attach the correct next amino acid to the growing strand of protein.

A key part of the system is a set of helper molecules called transfer RNA (or tRNA) that move the amino acids into place during the assembly of a protein molecule. In each type of tRNA molecule, one end sticks to a specific amino acid, while the other end sticks to a matching codon in the messenger RNA (mRNA). Once the amino acid is maneuvered into position by the tRNA molecule, the tRNA is ejected, and the

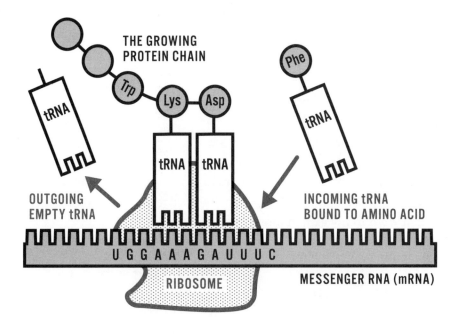

THE GROWING PROTEIN CHAIN

Phe

Trp Lys Asp

tRNA

tRNA

tRNA tRNA

OUTGOING EMPTY tRNA

INCOMING tRNA BOUND TO AMINO ACID

U G G A A A G A U U U C

RIBOSOME

MESSENGER RNA (mRNA)

How a ribosome constructs a protein molecule

ribosome moves on to the next codon, as shown in the illustration above. In this illustration, each three-letter abbreviation represents a specific amino acid, such as *Phe* for *phenylalanine*. RNA uses the nucleotide *uracil* (abbreviated *U*) in places where DNA would use thymine, which is why the letter *U* appears in the message carried by the mRNA.

If a gene is like a list of ingredients, could we also say a gene is like a recipe and DNA is like a cookbook? This analogy is far more apt than comparing DNA to a blueprint or a computer program, but it falls apart in a couple of ways. The first way is that a recipe has two parts: a list of ingredients and a set of instructions for making something from those ingredients (in other words, a step-by-step process). DNA lacks that set of instructions. The second way the analogy breaks down is that unlike a recipe, DNA contains no information about the quantity of each ingredient, such as 3 cups of flour or 1 teaspoon of baking powder. If a particular protein molecule needs to include twenty-six instances of the amino acid cysteine, codons for cysteine must appear twenty-six times in the gene that codes for that protein.

In a cookbook recipe, you primarily rely on the instructions to tell you the sequence for adding the ingredients, which may differ from the sequence of ingredients in the list. But in DNA, all you have is a list of ingredients; thus, the sequence of the codons is crucially important, because this controls the sequence in which the amino acids are strung together into a chain.

So while we might informally say that a gene contains the recipe for making a protein, it would be more accurate to say that a gene lists the sequence of amino acids for making a protein. However, because it's a lot shorter and simpler to use the word *recipe*, I'll continue to use that word in my explanations!

What Makes It All Work?

So how does DNA actually work? How does a collection of recipes for making proteins produce a human being (or an elephant, or an oak tree)? It seems intuitive that something must be controlling the entire operation.

You might think of the human body as a huge construction project, with each living cell serving as a job site within the project. The protein molecules are the workers, and each worker has one of ninety thousand specialties (the various types of proteins). Each cell hires its own workers and manages its own job site, but messages constantly go back and forth between the cells to help coordinate it all. Who directs the whole enterprise? Who sits at the top, making sure everything is going according to plan? (Hint: it's not your brain. Your brain controls a lot of things, but it does not control your development from a zygote to an adult, nor does it control the metabolic processes going on inside each cell.)

The best way to answer this question is to take a closer look at gene expression. At any given time, some of your genes are being expressed—copied into messenger RNA for the creation of new protein molecules—and others are not. As your body grows from a zygote into a fully developed human being, these differences in gene expression allow the cells to differentiate into various types, becoming skin, bone, muscle, blood, nerves, and so on. The differences in gene expression allow distinct organs to form within your body—heart, lungs, brain, liver, intestines, and so on. This grand project ultimately results in the final shape of your body, with two legs, five fingers on each hand, twelve pairs of ribs, thirty-three vertebrae, and so on.

Every cell in your body has its own copy of your DNA, and in each cell, gene expression operates independently, managed locally by the cell nucleus. Even for cells of the same kind, their specific location in the body (and many other factors) can result in a different expression of the genes. As a result, the management of this enterprise is highly decentralized. Each cell runs its own operation, doing what it needs to do every moment of its life in order to fulfill its specific role.

This amazingly intricate process involves a huge number of distinct steps, along with a lot of looping and branching, just like an algorithm in an enormously complex computer program. Theoretically, one could imagine such an algorithm encoded somewhere in the body, directing all the processes of the body. But imagine what a beast that algorithm would have to be in order to turn on and off each of the twenty thousand genes in each of the thirty trillion cells in your body based on an enormously long and intricate set of factors. Such an algorithm would be hopelessly complex and far too long to store anywhere, regardless of the encoding system. And imagine the communication logistics of trying to send messages to each of twenty thousand genes in each of thirty trillion cells. Just getting all of those messages to the right places, and in a timely manner, would be hopelessly complex.

The upshot is that the human body contains no central authority to micromanage gene expression across all the cells. Instead, it all works as an *emergent system*. The details of the human body emerge as a result of all the behaviors of the contributing parts. It's a bit like an ant colony, which functions as a coherent whole even though no central authority—not even the queen of the colony—micromanages the tasks performed by the individual ants. However, the development and maintenance of the human body is incredibly more complex than the behavior of an ant colony.

Given that each of the thirty trillion cells in the human body handles the process independently, what controls gene expression at the cellular level? What determines which proteins actually get made, when they get made, and in which cells they get made? This is a complicated topic, and there is much we don't yet know or understand, but scientific research is proceeding at a breakneck pace to bring many of the key puzzle pieces into view. We know that RNA molecules copied from certain areas in your DNA that don't code for proteins help to regulate gene expression. Protein-encoding regulatory genes also help to turn other genes on and off. Still other kinds of molecules attach themselves to genes and affect their expression.

A closely related issue is that cells need to cooperate with each other, especially with cells that are nearby. Lots of messages go back and forth between the cells, and most of these messages take the form of molecules. (Certain other messages are carried by electrical impulses in your nervous system.) Unlike messenger RNA, these molecular messages aren't based on a data encoding system and therefore don't consist of distinct words, comparable to codons. Instead, each signaling or regulatory molecule simply hooks onto a receptor molecule on or within the target cell, which triggers a reaction, causing a particular biochemical process to begin or end, or to go faster or slower.

Implications for Genetic Engineering

Let's review a few of the key ideas we've covered so that we can consider the practical implications of those ideas.

DNA IS NOT LIKE A BLUEPRINT. DNA does not contain any plans for the structure or appearance of an organism. If you take a single gene from a fish and insert it into an embryonic tomato plant, you won't get a plant bearing fish organs, such as fins or eyes. If the inserted gene is expressed at all, one additional protein will appear in some of the plant cells. This protein will probably have little effect on the plant. However, if the protein interacts significantly with other molecules, it could have a major impact on the plant

DNA IS NOT LIKE A COMPUTER PROGRAM. Human DNA does not contain any instructions for building or maintaining a human body. And while the processes that occur in the body can be compared to algorithms, DNA does not contain any information regarding any of the steps in those algorithms (although the protein recipes encoded in the DNA indirectly cause all of those processes to happen). Therefore, you cannot approach the editing of DNA in the same manner as you would approach the creation and editing of a computer program

DNA CONTROLS THE DEVELOPMENT AND MAINTENANCE OF THE BODY INDIRECTLY, THROUGH THE PROTEINS THAT IT ENCODES. Proteins drive most of the processes in the human body, and DNA contains the recipes for those proteins. Thus, DNA indirectly controls the development and maintenance of your body, although in a manner that's far from straightforward. To understand how the human genome controls these processes, you need to know at least four key things: (1) the proteins that are produced by each gene; (2) the molecules and processes that regulate the expression of each gene; (3) how each type of protein interacts with other molecules in the body, especially when those interactions mediate biochemical processes; and (4) how the products of multiple genes interact with each other to achieve various end results.

By studying DNA and the proteins it encodes for, we can draw some correlations between the presence of certain genes and the effects of those genes. Because of such correlations, scientists can use the DNA of an unknown person to predict the eye color of that person with a reasonable degree of certainty. But more often than not, the effects of our genes are quite complicated and depend on the interactions of the products of many different genes, combined with various environmental factors.

NO CENTRAL AUTHORITY, AND NO CENTRALIZED SET OF INSTRUCTIONS, CONTROLS THE DEVELOPMENT AND MAINTENANCE OF THE HUMAN BODY. Your body has trillions of cells, and every cell has its own copy of your DNA, including all twenty thousand genes. Every cell independently regulates the expression of those genes. This gene expression is influenced by the molecules that come in contact with the outside surface of the cell membrane, but otherwise each cell does its own thing. There is no central authority—not in the brain, nor in any other part of the body—that drives the development and maintenance of the human body. Instead, the overall results are an example of an emergent system, in which the behaviors of all the component parts result in a distinct set of characteristics and behaviors at a larger scale.

These key ideas have major implications for what could feasibly be accomplished with genetic engineering, before even getting into the moral and political issues. If DNA were like a blueprint or a computer program, people would be able to invent a completely new organism and to create the DNA necessary to produce that organism. Instead, the only viable approach is to start with the complete genome of an existing organism and then to make very small modifications to the DNA. This is, in fact, how genetic engineering is currently conducted.

Scientists take an existing genome—for example, that of a tomato plant—and then attempt to add or replace a single gene or a very small number of genes. A typical objective is to improve a single characteristic, such as cold hardiness or pest resistance. (Quite often, the inserted genes are found naturally in other members of the same species or in closely related species.) If and when the objective is achieved, the people conducting the research might try to improve another characteristic. It's a gradual and incremental process, just like traditional plant breeding and evolution due to natural selection. Even so, there are good reasons for society to engage in a serious debate about what specific limits to place on genetic engineering.

Order as an Emergent Property

As a person who appreciates a well-ordered system, I have great admiration for the many levels of organization in a single living creature, such as a human being. These levels include cells, tissues, organs, and more. So I was dumbfounded when I first realized that none of this organization is specified in our DNA—that our genes do not serve as a blueprint for our bodies. In fact, I was quite frustrated with this realization. If the structure of the human body is not explicitly detailed in our DNA, how does this organization come about? The answer, I learned, is not through any centralized system of control but through the attributes and interactions of the component parts at the molecular and cellular levels. I found this realization to be even more frustrating because I could not at first make any sense of it.

But over time, I gradually became aware that the world is chock-full of such emergent properties. In fact, I now see nearly everything around me with this lens. For example, how do we explain the properties of water? These properties emerge as a result of the properties of water molecules, which arise from the properties of the constituent atoms, which in turn arise from the properties of the participating atomic particles (protons, neutrons, and electrons), which in turn arise from the properties of the constituent quarks. At a much larger scale, the amazing shapes we see in a flying flock of starlings or a swimming school of mackerel are also an emergent property—the result of the independent actions of the individual animals. Those are just two examples, at very different scales, but the examples are countless, and the human body alone contains a great many examples. These examples occur at many different scales, ranging from atoms to molecules to cells to tissues to organs to organ systems (such as the circulatory system). At each level of this hierarchy, new properties emerge as a result of the interactions of the constituent parts one level down.

So now, instead of feeling frustrated, I'm pleased. It seems that nature, like me, has an obsessive-compulsive tendency to impose order on the world, creating organization out of chaos. I'm really proud of you, Mother Nature! Keep up the good work!

9

Superfoods and Toxins

I typically cook dinner from scratch three times a week, starting mostly with fresh ingredients, including lots of vegetables and a little bit of meat. A majority of my recipes are one-pot meals, in which the meat, vegetables, and other ingredients are all eventually combined in a pot, pan, or casserole dish. I cook a large quantity each time, enough to provide two people with four meals (two lunches and two suppers). It usually takes me a couple of hours to prepare the meal, in part because I insist on cutting up all the raw ingredients into bite-sized pieces. People who see my slow, methodical chopping find it either amusing or maddening, but since I'm usually alone in the kitchen, the process gives me time to think.

Not surprisingly, I often think about food while I'm cooking. There's plenty to think about. Few topics are as fascinating to the general public as food and nutrition, and few aspects of life offer such profit potential for entrepreneurs who successfully hitch a ride on the latest fad. For these reasons, strange new concepts about food and nutrition are generated and promoted at a furious pace. Hiding in all this noise is a small dribble of genuine scientific information. Unfortunately, it's hard to distinguish these nuggets of truth from the many popular ideas that sound perfectly legitimate but are in fact misleading. Two overused buzzwords—*superfoods* and *toxins*—are especially in need of scrutiny. Most superfoods are indeed quite nutritious, but articles in the media describe these foods as if they had magic powers. Even more misleading is the promotion of foods or treatments to "cleanse the toxins from your system," completely distorting the concept of toxins and how the body actually gets rid of them.

It makes perfect sense to pursue the goal of a healthy diet, as the types and quantities of foods you eat can certainly have a significant effect on your health. However, the nature of that impact has changed over time as the foods available and our eating habits have both evolved. Many Western countries today (particularly the United States) are plagued with diabetes and heart disease, both of which are often associated with eating too much of the wrong foods, combined with not getting enough exercise. A century ago, these two diseases were relatively rare. Our biggest diet-related health issues were malnutrition and vitamin deficiency, typically related to a shortage of food, or of certain kinds of food. These ancient problems have not completely disappeared, but in many parts of the world the details have changed. Today it's easy to be overweight from overeating and yet suffer from malnutrition due to eating a highly unbalanced diet.

It seems that most people have internalized the general concept that some foods tend to be health promoting while other foods tend to be health degrading. While this mental model has significant truth to it, it's also a huge oversimplification because quantities must be taken into account as well. Foods that are harmless or even beneficial in small quantities can become harmful in large quantities. But the main difficulty with this model is the endless noise in the popular media regarding which foods are "good" and which are "bad." This ubiquitous (and dubious) advice generates an endless series of food and diet fads, making it difficult for the average person to separate fact from fiction.

The Concept of Superfoods

One of the most popular buzzwords regarding food and nutrition is *superfood*. Every few years articles in the media aggressively promote a newly designated superfood. The typical approach is to focus on a relatively little-used fruit, vegetable, or grain and to suggest that recent discoveries have shown that eating this particular item will make dramatic improvements to your health. The popularity of that food item will usually skyrocket (along with the price), and new packaged foods will appear that contain small amounts of the loudly promoted super-ingredient. Some of the many items that have been promoted as superfoods include pomegranates, quinoa, acai, kale, blueberries, salmon, seaweed, chia seeds, goji berries, oatmeal, garlic, green tea, macadamia nuts, and dark chocolate.

Most of these items are indeed wonderful foods, but it's important to note that the word *superfood* is not a scientific term; it's a marketing term. The word has no scientific definition, and the writers of popular media articles (not scientists) are the ones who decide which foods to bestow the term upon. These writers often attempt to slap a veneer of science on each designation by claiming that "recent studies" have proven how incredibly healthful a particular superfood is. In reality, the limited studies that do exist seldom support these outsized claims.

While most superfoods are indeed nutritious and are potentially useful additions to a balanced diet, they aren't magic. Including a few of these foods in your diet is unlikely to have a huge effect on your health. A superfood earns the designation by having an elevated level of certain beneficial nutrients. For example, quinoa is an excellent source of plant-based protein (with a good balance of amino acids), while blueberries are an excellent source of antioxidants. But you can get amino acids and antioxidants from a great number of other sources, some of which are seldom mentioned as superfoods. To be clear, if your current diet is dangerously low in certain essential nutrients, consuming a greater quantity of those nutrients can indeed have a significant effect on your health. But if you've been eating a balanced diet, with plenty of fresh ingredients, you're probably not suffering from any extreme nutrient deficiencies.

This brings up an important point: more than one mental model can be used to explain how superfoods work. In one popular model, a superfood is like a magic potion that confers generic "good health" on the fortunate recipients. A far more nuanced model—presented in the previous paragraph—suggests that superfoods can help you obtain important nutrients that ought to be in everyone's diet. However, the likelihood of experiencing a benefit is much greater if you start with the natural, whole version of that food rather than relying on a packaged product that claims to include the superfood as one of its ingredients.

Relying on packaged, processed products as your source of superfoods has three principal problems:

1. As a general rule, packaged foods typically include only very small amounts of the promised superfood ingredient.
2. Packaged foods often include large quantities of ingredients that should not be consumed in such large quantities, such as sugar, fat, starch, and salt.
3. The key nutrient the superfood is supposed to deliver might not be present anymore, having been lost or destroyed during the processing of the food.

For example, a beverage that claims to contain acai might in fact consist primarily of cheaper juices such as grape and apple, with only a trace of acai. That same beverage is likely to include large amounts of sugar. Even though the sugar might be derived entirely from natural sources (such as supersweet varieties of grapes and apples), drinking large amounts of any type of sugar is dangerous to your health. And there is little assurance that the nutritional value of the original acai berries is still present after all the processing. Some of the value is lost when the crushed pulp is discarded. Certain other nutrients are sensitive to heat or light and degrade during the processing, possibly diminishing further during the period between manufacturing and consumption.

What Are Nutrients, Anyway?

This raises a key question: What exactly does the word *nutrient* mean? A nutrient is any naturally occurring chemical compound found in food that the human body needs on a continuous basis in order to function normally. We usually divide nutrients into five principal categories: proteins, fats, carbohydrates, vitamins, and minerals. We typically need proteins, fats, and carbohydrates in much larger quantities than we do vitamins and minerals, but that doesn't make vitamins and minerals any less essential.

VITAMINS AND MINERALS

Because of the small quantities needed, it was only about a hundred years ago that we finally became aware of vitamins as a result of several medical mysteries. For example, scientists realized that something in butter could sharply reduce the rate of childhood night blindness and other eye diseases, but they didn't know which component of butter was responsible. They simply called it vitamin A. Scientists were also aware that something in citrus fruits could prevent scurvy. Again, they didn't know what it was but called it vitamin C. Likewise, vitamin D can prevent rickets, and vitamin B (which is actually a complex of perhaps eight different compounds) can prevent pellagra, beriberi, and other conditions.

Over time, scientists figured out the exact chemical compounds for all of the vitamins. Vitamin A is retinol, derived from beta-carotene in our food. The eight recognized B vitamins include niacin, thiamin, riboflavin, and folic acid. Vitamin C is ascorbic acid. Vitamin D is a set of three closely related calciferol compounds, designated as D_1, D_2, and D_3. Vitamin E is a set of several closely related tocopherol compounds, particularly alpha-tocopherol. Vitamin K, which is essential for blood clotting, comes in two main forms, phylloquinone and menadione, designated as K_1 and K_2.

So what exactly is a vitamin? A vitamin is a carbon-based chemical compound (in other words, an *organic* molecule) needed continuously in small quantities by the human body but that the body is unable to synthesize on its own. It's noteworthy that most animals can synthesize Vitamin C in their bodies but humans and other great apes lost this ability due to a genetic change in our ancestors.

Minerals are simple *inorganic* molecules and atoms needed by the human body in small quantities in order to maintain good health. Most of these minerals are metals, and we usually express our nutritional needs in terms of the quantity of the pure element, although the minerals we actually consume are usually bound up with other atoms. Other than water and salt (sodium chloride), the minerals we need in the greatest quantities are potassium, phosphorus, calcium, and magnesium. In much smaller quantities, we also need iron, zinc, manganese, and copper. In still smaller quantities, we need iodine, selenium, molybdenum, and chromium. Each of these elements plays a specific role (or sometimes multiple roles) in the human body. For example, an atom of iron lies at the center of each molecule of hemoglobin in your red blood cells, an atom that's crucial for enabling your red blood cells to transport oxygen around the body. If you fail to get enough iron, you may suffer from anemia due to your body having insufficient hemoglobin for building new red blood cells.

The US government and other institutions have long attempted to specify the minimum amount of each vitamin and mineral that should be present in your daily diet in order to avoid nutritional deficiencies. Consuming somewhat larger quantities might be beneficial, but there's considerable debate as to the *ideal* daily consumption of each vitamin and mineral, as opposed to the *minimum* amount needed to avoid serious deficiency diseases. Some vitamins and minerals are toxic when consumed in excess, an important factor that should be taken into account. Still, the ideal daily intake for most vitamins and minerals is likely to be higher than the officially recommended minimum amount. Also note that the ideal amount can vary according to the age and gender of the person, along with the long-term and short-term conditions specifically affecting that person.

PROTEINS, FATS, AND CARBOHYDRATES

Because vitamins and minerals are needed in such small quantities, they are sometimes called micronutrients. The three types of nutrients we need in much larger quantities are proteins, fats, and carbohydrates, all of which are complex molecules that must be broken apart into simpler molecules before they can pass from your digestive system into your bloodstream. (The simplest sugars, such as glucose and fructose, are exceptions and can be absorbed exactly as they are.) Your body treats proteins, fats, and carbohydrates much like LEGO blocks, first disassembling them into their core components, then later reassembling the components into whatever configurations are currently needed.

Each protein molecule consists of a long string of simpler molecules called amino acids. Your digestive system separates the twenty different kinds of amino acids from one another, which allows them to be absorbed into the bloodstream. Your body uses these amino acids to build the thousands of different proteins specified by the "recipes" in your DNA, so it's crucial that your bloodstream contain enough of all the amino acids. Your body is capable of synthesizing eleven of these amino acids, which leaves nine "essential" amino acids that must be present in your food. The best sources of complete proteins (containing all nine essential amino acids) tend to be meat, dairy, and eggs. Because many plant-based foods are deficient in certain essential amino acids, a person on a vegan diet must be more alert to the task of consuming a balanced set of proteins. For example, rice and beans are each deficient in certain amino acids, but a mixture of rice and beans is balanced. An even simpler solution is to eat quinoa, soy, or buckwheat, each of which provides a balanced, complete mix of amino acids. Thus, even vegans have multiple ways to obtain all of the essential amino acids.

Fat molecules are also disassembled by your digestive system. Each fat molecule consists of four parts: three fatty acids and a tiny glycerin molecule. There are many kinds of fatty acids, so the challenge is to get a proper mix of them in your diet. Some fatty acids, such as the ones called saturated fats, which contain no double bonds in their carbon

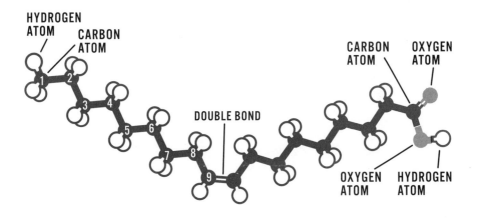

Oleic acid, which has just one double bond, located nine carbon atoms from the end, making it an omega-9 fatty acid

chains, are thought to be harmful when eaten in large quantities. Other fatty acids are considered to be essential for maintaining good health. We differentiate the fatty acids according to the length of their carbon chains and by the quantity and location of any carbon-to-carbon double bonds in the chain. A monounsaturated fatty acid, such as the oleic acid found in olive oil, has exactly one double bond somewhere in the chain. A polyunsaturated fatty acid, such as the linoleic acid also found in olive oil, has more than one double bond in the chain. When the term *omega* is used to categorize a fatty acid, the term refers to the position of the first double bond in the chain, starting from the end that has no oxygen atoms. For example, oleic acid is an omega-9 fatty acid.

Carbohydrates form the third category of food molecules that must be disassembled by your digestive system. The simplest carbohydrates are the simple sugars mentioned earlier, such as glucose and fructose. These are often called monosaccharides, meaning that they consist of a single looped chain that includes six carbon atoms. Compound sugars, such as sucrose (table sugar), consist of two such loops linked together and hence are called disaccharides. Any digestible carbohydrate with more than two saccharide units is a starch, also called a polysaccharide. Your digestive system tears apart the carbohydrates into individual saccharide units, which are then absorbed into the bloodstream. Monosaccharides in the blood are used in the construction of other molecules needed by the body, but the main use of "blood sugar" is to provide energy to all the cells of the body.

Other Important Components of Foods

Beyond the five traditional categories of nutrients—vitamins, minerals, proteins, fats, and carbohydrates—food contains other components that can be beneficial to human health. Perhaps the most obvious such component is water. Humans need a lot of water to remain healthy. Technically, we could call water a mineral, but it's so different from the other minerals—and we need it in such large quantities—that it deserves its own category.

Another crucial component of food is fiber, although this particular component seldom gets enough attention. We get our fiber from plant-based foods—fruits, vegetables, and grains. The word *fiber* is highly misleading, with a completely different meaning from other contexts in which the word is used, such as textile manufacturing. Dietary fiber is essentially anything that passes through your digestive system without getting digested. Protein fibers are not dietary fiber, because protein is digestible. But any form of cellulose, which is found in all plants, is fiber, because humans cannot digest cellulose.

It seems odd that something you cannot digest could be an important part of your diet, but fiber promotes health in several different ways. First, fiber normalizes bowel movements, reducing the chance of constipation while also reducing the odds of loose, watery stools. Fiber assists in moving waste through the intestines, preventing the waste from remaining in your bowels for too long, where it can do harm. Because fiber binds with other substances, it reduces the amount of cholesterol that enters your blood, and it smooths out the absorption of sugar, which helps you to maintain consistent blood sugar levels. (In particular, consuming fresh fruit instead of fruit juice slows down the absorption of sugars so that the sugar does not enter your bloodstream too quickly.) A diet rich in fiber tends to have fewer calories than a diet low in fiber, making it easier to control one's weight.

I once thought dietary fiber consisted entirely of actual fibers, like the stringy material found in overmature asparagus or okra. But in reality, most dietary fiber does not take the form of visible fibers. In fact, a lot of dietary fiber is water soluble and becomes a gel. For maximum health benefits, you should have generous amounts of both soluble and insoluble fiber in your diet. Soluble fiber is found in oats, peas, beans, apples, citrus fruits, carrots, barley, and psyllium. (Pectin, a compound found in apples and other fruits, is a good example of soluble fiber.) Insoluble fiber is found in whole wheat flour, wheat bran, nuts, beans, and many kinds of vegetables, including cauliflower, green beans, and potatoes.

Antioxidants are another important component of food. Several superfoods (such as blueberries and acai) have achieved that designation primarily due to the presence of antioxidants. These are chemical compounds that tend to react with free radicals, thereby neutralizing them.

But what in the world are free radicals? The underlying concept is that many types of chemical reactions involve an atom or molecule giving up an electron to some other atom or molecule that accepts the donated electron. This binds the two pieces together in an ionic bond. Free radicals are unstable atoms or molecules that are highly prone to getting involved in such reactions. Some free radicals appear in your body as by-products of normal chemical reactions, such as the metabolism of sugars in the mitochondria. Other free radicals can appear as a result of external factors such as pollution.

Excessive quantities of free radicals in your body result in oxidative stress, producing undesirable or harmful reactions. Oxidative stress has been linked to a wide range of conditions, including heart disease, cancer, arthritis, stroke, respiratory diseases, immune deficiency, emphysema, and Parkinson's disease. Antioxidants in your food can reduce the quantity of free radicals in your cells and blood, thereby helping to keep you healthy. On the other hand, it would be wrong to assume that all free radicals are harmful to your body. Your immune system employs free radicals in its attacks against invaders, and your ability to extract energy from your food depends on the action of free radicals.

Antioxidants come in many different forms. The vitamins A, C, and E, along with the minerals selenium and manganese, all have antioxidant properties. Most of the other well-known antioxidants are found exclusively in plants and provide much of the color in fruits and some vegetables. Lycopene and lutein are two examples of antioxidants, along with various flavonoids, polyphenols, and other compounds. Foods with high levels of antioxidants include blueberries, tomatoes, carrots, oranges, bell peppers, spinach, kale, and watermelon.

It should be noted that some antioxidants, such as vitamin C, are destroyed by cooking. Thus, it's helpful to include some raw fruits and vegetables in your diet, in addition to the cooked vegetables (which have their own advantages).

The Magic Bullets of Medicine and Diet

These days, if you Google the phrase *magic bullet*, you'll get many pages of results that promote a certain brand of kitchen blender. However, when the phrase originated more than a century ago, it referred to the discovery and use of chemical compounds that have an amazing ability to target and cure diseases in the human body with few serious side effects. Many of our best examples of magic bullets are antibiotics. As a kid, I was able to witness this magic firsthand. My younger brother was hospitalized with pneumonia, which turned out to be bacterial in origin. He was given an antibiotic, and within a couple of days he was transformed from a dangerously sick and lethargic little child into a happy and healthy ball of energy running around the hospital floor. Today many of our antibiotics are becoming less effective due to overuse (which breeds antibiotic resistance in the target bacteria), but this broad category of drugs is still an outstanding example of magic bullets in medicine.

As a young man, I had an even more personal experience with a magic bullet. While traveling internationally, I contracted hepatitis A from the food or water. Several months later I was still bedridden and eventually hospitalized due to a complete blockage of my bile ducts, even though the virus had gradually disappeared from my system. My eyes and skin were yellow with jaundice, and my skin never stopped itching. Finally the doctors gave me some kind of cortisone drug, and within a day or two my bile ducts were back to normal and I could begin my recovery. I was astounded because I had been unaware that modern medicine provided us with any magic bullets other than antibiotics and vitamins.

Of course, antibiotics are considered to be drugs, while vitamins are not. Vitamins and drugs are both chemical compounds that are introduced into the body with the goal of having a beneficial effect on health. The difference is that your body needs vitamins on a daily basis to maintain its normal biochemical operations, while drugs are typically used to attack and correct abnormal conditions, such as an infection by harmful bacteria. But even though vitamins are not drugs, they can serve as magic bullets for treating people with serious vitamin deficiencies—that is, with conditions such as scurvy, rickets, beriberi, and pellagra. For people without serious vitamin deficiencies, taking vitamin supplements does not have such a dramatic effect, so vitamins are not magic bullets for these people. Still, it makes sense to ensure you're getting an adequate supply of vitamins in your diet.

The discovery of vitamins, antibiotics, and other magic bullet compounds has been incredibly beneficial for humankind. But a subtly harmful side effect has impacted our thinking. We are so accustomed to magic bullets that we expect every conceivable malady to have an easy solution. Whatever the problem, we expect there to be a drug or a food that will cure it—immediately, totally, and without side effects. Far from being complete cures, many of our most recent drugs have only a marginal effect on the target condition. A "successful" cancer drug might increase the life span of a typical cancer patient by only a few months. A "successful" antiviral drug might shorten the typical course of the disease by only a few days.

This brings us back to superfoods. These foods are not magic, and they are unlikely to cure all of your ailments, despite the frequent headlines in the popular media and the stories on the internet. On the other hand, even though the supposed benefits are highly exaggerated, these foods are not completely worthless. On the contrary, many of them are well worth including in your diet.

The Trouble with Talking About Toxins

Toxin, like superfood, is one of the dietary buzzwords found most frequently in the popular media. The word toxin, in contrast to the word superfood, is a genuine scientific term with a clearly defined scientific meaning. And yet much of what is reported in the popular media about toxins is utter nonsense. The concept that you need to actively "rid your body of toxins" is based on falsehoods and misunderstanding. Products that claim to "cleanse," "detox," or "flush" your body of these supposed toxins seldom do you any good—and in some cases can actually harm you. Furthermore, most of these advertisements and stories either fail to define the word toxin or else equate the word with man-made pollutants, which is not what the word actually means. It seems odd that this huge informational gulf should exist between science and the popular media, so let's take a look at what the word toxin means to scientists.

A toxin is any poisonous substance produced by a living organism. Scientists often use the word in the context of microbiology, referring to toxic compounds given off by certain kinds of bacteria. The most powerful toxin known to humankind is the botulinum toxin, produced by the bacterium Clostridium botulinum under anaerobic conditions (that is, in places where no oxygen is available, such as in jars of home-canned foods that were prepared without adequate sterilization). Certain bacterial diseases are harmful primarily due to the release of bacterial toxins rather than the direct effects of the bacteria on human cells. A prime example is tetanus, a disease caused when the bacterium *Clostridium tetani* grows in wounds (especially anaerobic wounds, such as punctures), releasing a powerful toxin into the human body. Interestingly, the vaccine for tetanus causes the body to produce antibodies that attack the toxin molecules rather than the bacterium, reflecting the primary importance of neutralizing the toxin. However, a person who experiences a risky wound might be given antibiotics or antibodies that attack the bacteria directly.

The word *toxin* can also apply to poisonous substances produced by animals for their defense or for hunting. Certain toads and frogs produce toxins in their skin, making them less desirable to predators. Venomous snakes produce a surprisingly diverse set of toxins in their venoms. Poisonous compounds in plants can also be called toxins; these include atropine and scopolamine in deadly nightshade, nicotine in tobacco, and the toxic alkaloids found in hemlock. Alcohol, a waste product produced by yeast cells, is also a toxin, and a person who has ingested a large but subfatal dose is said to be intoxicated.

Other kinds of waste products can also be toxic. Our own human cells produce waste that must be disposed of and that can become toxic if the quantities become excessive. When scientists speak of toxins in the human body, they are primarily referring to two things: (1) waste produced by your own cells, and (2) compounds released by microorganisms in your blood and in your guts. While it is true that certain toxic substances might enter your body from your food and drink, or from the air, or from contact with your skin, most of these compounds are not technically toxins.

In the popular media, the meaning of the word *toxin* has been expanded to include anything that could possibly be toxic. Stories and advertisements typically emphasize pollution and other by-products of modern industrial societies, thereby completely abandoning the original concept that toxins are the natural (but toxic) products of living organisms. A more appropriate term for environmental pollution would be *toxic compound* or *toxicant*. You do indeed have reason to be concerned about toxicants such as heavy metals (including mercury and lead) and certain synthetic organic compounds (such as those found in many pesticides).

On the other hand, the stories in the popular media often deviate considerably from the actual science. One misleading trick is to suggest that if a compound is toxic in high doses, extremely low doses must also be quite dangerous. This is not always the case, especially if the compound is quickly excreted or denatured by the body. A good example is formaldehyde, which is toxic in high doses but which is naturally produced by your own body and therefore is present in your blood in low concentrations at all times. And as already discussed, several types of vitamins are toxic in large quantities, even though small quantities are essential for health.

Cleansing Your Body of Toxins

In the popular media, a frequent theme is that you need to "cleanse your body" of toxins. In reality, the human body already has several well-developed mechanisms for removing toxins, and these cleansing mechanisms occur automatically on a continuous basis, without your conscious intervention. Your body employs four principal methods to deal with toxins and other toxic compounds:

1. detoxification by the liver
2. removal by the kidneys
3. evacuation by the large intestine
4. disabling by antibodies in the blood

DETOXIFICATION BY THE LIVER

The liver is a multifunctional organ that serves several crucial roles, one of which is the disabling and removal of toxins from the body. The liver filters a wide range of toxins from the blood and then processes these toxins in multiple ways. The processing typically changes the chemical forms of the toxins, producing modified substances that are less dangerous and easier to excrete from the body. For example, fat-soluble toxins are changed into water-soluble molecules. Some of the modified substances are then released back into the bloodstream for removal by the kidneys, while others are released into the bile for evacuation by the intestines.

The alcohol in alcoholic beverages is harmful to human cells, and therefore alcohol is one of the toxins that is detoxified by your liver. Excessive consumption of alcohol destroys liver cells, eventually leading to cirrhosis and possible liver failure. Certain other toxic compounds— such as carbon tetrachloride, a dangerous substance that was once commonly used for dry cleaning—can have similar effects on the liver.

Although some toxins can enter your body from the outside, it is helpful to remember that your body constantly generates its own toxins due to completely natural processes. Therefore, even if you avoided all external sources of toxic compounds, your liver would still have a full-time job removing the natural toxins produced by the cells of your body, the result of normal cellular metabolism and the breakdown of spent cells.

REMOVAL BY THE KIDNEYS

The kidneys specialize in removing water-soluble materials from the bloodstream, to be excreted from your body in the form of urine. (The kidneys also perform other duties.) Urine tends to be about 95 percent water, with the rest composed primarily of waste products and excess minerals. Other than water, the largest component in urine is usually urea, a yellow nitrogen-containing compound. This is how your body safely disposes of ammonia, a waste product generated from the processing of spent proteins and other nitrogenous compounds. Urea is produced in the liver by combining waste ammonia with carbon dioxide, which neutralizes the toxic ammonia molecules. The kidneys also remove other toxins that have been processed by the liver and returned to the blood.

EVACUATION BY THE LARGE INTESTINE

After the food you have eaten has been digested into nutrients that can be absorbed into the bloodstream (such as amino acids, fatty acids, and simple sugars), most of that absorption occurs in the small intestine. The large intestine then prepares the food waste for evacuation, primarily by absorbing excess water. Toxins in this waste come from two primary sources: bile from the liver, and the action of microorganisms in the intestines. However, both of these sources have positive roles as well.

The bile produced by the liver contains more than just the processed remains of toxic waste. Bile contains enzymes that are important for digesting the fats in foods, breaking those fats down into fatty acids. Just as the liver is a multifunctional organ, the bile produced by the liver is a multifunctional fluid. Although the waste included in the bile has been made safer by the liver's processing, not all of that waste is completely safe, and it's better if the waste moves through the large intestine in a timely fashion.

A wide variety of microorganisms live in the intestines, and some of those organisms help release additional nutrients from the digested food. For example, perhaps twenty different species of bacteria play a role in the production of eight different B vitamins in both the large and small intestines. Bacteria also play an important role in the production of vitamin K in the large intestine. At the same time, certain other bacteria in the intestines release toxins as a result of their normal activity. So again, it is better if the waste moves through the large intestine in a timely fashion. Toxins that remain in the large intestine might be absorbed (or reabsorbed) into the bloodstream, or they might cause harm to the large intestine itself.

The best way to ensure that toxins are quickly removed from the large intestine is to eat a high-fiber diet. Eating plenty of vegetables, whole fruit (not fruit juice), and whole grains is a great way to ensure plenty of fiber in the diet, in addition to being a source of vitamins, minerals, and other nutrients. A diet high in fiber is strongly associated with a reduced risk of colon cancer, in large part due to the efficient removal of toxins from the large intestine.

DISABLING BY ANTIBODIES IN THE BLOOD

Anything that triggers a response by antibodies in the blood is called an antigen. In some cases an antigen can be an entire virus or bacterium, but quite often an antigen is simply a free-floating molecule, such as a toxin. In fact, when antibodies attack a virus or a bacterium, they are actually attacking specific protein molecules on the surface of that microorganism. So while we often associate antibodies with their role in controlling invasive germs, antibodies also play an important role in neutralizing and removing certain toxins in the body, especially toxins produced by infectious disease organisms.

The upshot of this discussion is that your body is quite adept at removing toxins, a task it performs continuously every day of your life. Therefore, it's highly misleading to tell people that they should "cleanse the toxins" from their systems, as if your body were unable to perform this task without your conscious intervention. Most of the heavily promoted "detox" products and techniques have little or no value. Simply eating a well-balanced diet with a lot of fiber and a minimum of processed foods is probably the best way to tune your body for maximum efficiency in the removal of toxins, toxicants, and other waste.

A Healthful, Balanced Diet

Today, as I finish chopping up the vegetables for my evening cooking, I look across the kitchen countertop at the neat bowls of evenly sized pieces of zucchini, carrots, mushrooms, celery, cabbage, red bell pepper, and onion. This seems like a fitting time to wrap up my thoughts about food and nutrition.

Food is an exceedingly popular topic in the media. New ideas and suggestions appear and spread at enormous speeds, generally falling into two categories. The first category emphasizes the enjoyment of food—preparing or purchasing food that is a pleasure to eat. The second category emphasizes a healthy diet—in other words, eating food that supports a person's health rather than degrading it. Quite often, these two categories overlap, as people seek to prepare and consume food that's simultaneously healthful and tasty. I'm certainly on board with that! The challenge when you encounter such information is to sort the wheat from the chaff, identifying the nuggets of good scientific information hidden in the flood of misinformation and hype. This is more easily said than done, but with practice we can all get better at it.

The good news is that the concept of superfoods might be evolving into something more realistic. These days, if you do a Google search of the term, the results will include a few truly informative articles about superfoods written by people who actually understand the science of nutrition. Rather than emphasizing obscure and trendy foods, these articles tell us that the *real* superfoods are berries (of all sorts), fish, leafy greens, nuts and seeds, olive oil, whole grains, active-culture yogurt, cruciferous vegetables (such as broccoli), and legumes (beans and peas). Other outstanding foods include avocados, tomatoes, garlic, mushrooms, eggs, sweet potatoes, oats, and brewed tea, plus quite a few others. It's true that certain rare and expensive foods can also be good for you, but they aren't any better than these commonplace examples. The underlying message is to eat a wide range of everyday superfoods—purchased fresh and whole rather than as processed foods, pills, or powders—as the major constituents of a healthful, balanced diet.

Okay, now it's time for me to start cooking with those beautiful chopped-up vegetables.

10

Full of Energy

Except when I'm traveling, I prefer to cook dinner rather than eat out. I keep a list of about thirty-six active dinner recipes, along with a list of currently inactive recipes. Each time I cook, I choose a recipe near the top of the active list and then move that item to the bottom of the list. It takes several weeks for any recipe to work its way back up to the top of the list, ensuring a diversity of dinner menus. Sometimes I decide to cook something new—something not in my usual repertoire that I've heard about or eaten in a restaurant. In that case, I feel a need to gain an intuitive understanding of the dish before undertaking the venture.

The very first time I prepare any new dish—whether minestrone, ratatouille, stuffed peppers, chicken gumbo, pad thai, avgolemono, German potato salad, shakshuka, tamale pie, tom kha gai, or even pumpkin pie—the first thing I do is to find at least three separate recipes for that dish, and preferably four or five distinct recipes. Then I carefully compare the recipes in a spreadsheet to find the commonalities and differences among them. This tells me which aspects of the dish are mandatory and which are simply variations. Finally, I write down a new, original version of the recipe, combining the core essence of the dish with my own touches, such as including additional vegetables. In short, I feel like I need to understand the dish at an abstract level before I can tackle any concrete implementation of the concept.

Admittedly, in the context of cooking, my process may sound a bit extreme. However, this way of thinking permeates much of what I do. Whenever I'm faced with an issue, rather than promptly diving into the details, I first need to understand the big picture to gain an intuitive feel for the abstract essence of the matter at hand. This approach strongly influences my thinking about natural science. For any science topic I encounter, I insist on understanding the big picture before studying any of the details. Details are useless to me—nothing more than trivia—unless I can see how they contribute to a bigger picture. But once I understand the big picture, the details become fascinating, and I soon become an avid collector of pertinent details.

Although this approach has served me well throughout my life, it has only led to frustration when it comes to the topic of energy. From fourth grade on, I could make little sense of the various definitions of the word *energy* in my textbooks. Because I lacked a clear and comprehensive mental construct for the meaning of energy, I found it hard to connect the many diverse details. To this day, I find it maddeningly difficult to provide a simple and intuitive definition of energy that applies equally to all forms of energy. The most popular definition—"the capacity for doing work"—just doesn't work for me.

So What Is Energy?

Richard Feynman, the Nobel Prize–winning physicist, discussed the law of conservation of energy in one of his famous lectures. Up front, he pointed out that *energy* is a highly abstract idea—essentially a mathematical concept—when the term is applied across all the many forms of energy. It is only when we talk about a specific form of energy that we can provide concrete physical details. In other words, the concept is notoriously difficult to describe in a comprehensive yet understandable manner. Yet energy is a very useful concept, and you've got to understand energy to make sense of nature and the world around us.

To grasp the concept of energy, don't start with the big picture; start by understanding the various forms of energy, one by one. Each form of energy has its own distinctive big picture. The key unifying concept is that energy can never be destroyed; it can only change form. Furthermore, the importance of energy is almost always connected to the situations in which it changes form. Consequently, the most important questions tend to be these:

- Is the energy in this system increasing or decreasing?
- Where is the energy coming from or going to?
- How much energy is coming or going?

Of course, you probably don't stay up late at night worrying about a universal definition for *energy*. Instead, you've developed your own intuitive understanding of energy based on several distinct contexts. You might talk about energy production or the energy industry. You might talk about your energy bill, and how it goes up in winter and again in summer. These contexts are connected to our society's infrastructure for distributing energy—mostly in the forms of electricity and fossil fuels—and how we depend on this infrastructure for powering our transportation and our machinery, and for heating our homes and offices.

And then there is the completely different context in which we associate the concept of energy with our bodies. We might say that we're "full of energy" as we launch into an important task but "out of energy" after having worked on that task. We might think of sugary food as a source of "quick energy" but also recognize the need to "burn off energy." In other words, we recognize that our bodies need energy and we get this energy from our food, but that a balance between eating (taking in energy) and exercise (using up that energy) ought to be maintained.

And finally, there is a third, more mystical way some people use the word *energy*. They might use phrases such as "life energy" or "negative energy" to talk about their experiences or their interpretation of the world around them, even though these terms have no meaning to a scientist. So how do we distinguish between uses of the word *energy* that have a scientific meaning and those that do not? Let's start by itemizing the principal forms of energy that are indeed recognized by science.

Light / Radiant Energy

The terms *light, radiant energy*, and *electromagnetic radiation* all refer to essentially the same thing: a form of pure energy that travels at the speed of light. We feel comfortable with the concept of light because our eyes detect light, allowing us to see the world around us. However, we often forget that our eyes detect only a very limited range of wavelengths. Any wavelength of light that's longer or shorter is invisible to us, including radio waves, microwaves, infrared radiation, ultraviolet radiation, X-rays, and gamma rays.

Light consists of photons, which are to light what atoms are to matter. In other words, light cannot be divided into a unit any smaller than a photon. It seems intuitive that light should be infinitely divisible—that no matter how small the quantity, we should be able to divide that quantity in half. But our intuition tells us the same thing about matter, and it isn't true there either. The indivisibility of photons leads to all sorts of interesting phenomena, including the properties of pigments and our ability to see colors.

Light doesn't need a medium to travel through; it's perfectly capable of traveling through the vacuum of space. Its rate of travel is incredibly swift; at roughly 186,000 miles per second, it covers a distance equal to the circumference of the earth in one-seventh of a second. We don't often think about this incredible speed when we sit in a room illuminated by an electric lamp. However, all the light that emerges from that lamp comes and goes in a microsecond. New photons of light constantly emerge from the lamp, replacing those that have just disappeared, producing an illusion of constancy.

It's amazing to consider that each of those photons is in our presence for less than a microsecond. In fact, if we're in an enclosed room, with heavy curtains drawn—thereby blocking any route where light might escape—each of those photons *exists* for less than a microsecond. They are generated by the lamp, travel at the speed of light until they strike an object, and then are either absorbed or reflected. If reflected, they repeat the process until they are absorbed by something in the room. Oddly, you cannot detect any of these photons while they are in motion. Your ability to detect light completely depends on your body absorbing some of the photons that strike you—in the retina, which enables you to see, or in the skin, providing warmth. And once your body absorbs a photon, that photon no longer exists; it has changed into a different form of energy.

This example illustrates the concept that the importance of energy is almost always associated with a change of form. The photons emerge from your lamp because electrical energy is converted to light. The photons promptly disappear when they are absorbed by matter, and this injects heat energy into the matter.

Heat Energy

We all have an intuitive sense of the concept of heat, although these popular ideas sometimes differ from the scientific concept of heat. Heat is a property of matter, a result of the movement of individual molecules within that matter. Molecules in matter are always in motion, either bouncing around freely (in a gas), sliding around (in a liquid), or vibrating in place (in a solid). With our human senses, we cannot detect individual molecules—much less the movements of these molecules—but we can certainly detect the net effect of their motion, which we perceive as heat. If an object feels hot to your touch, it's because some of that molecular movement is transferred from the hot object into the molecules of your skin.

We all know that the concept of temperature is associated with heat, although we might incorrectly assume that temperature is a measure of heat energy. In reality, temperature is a measure of how quickly the molecules in a mass are currently moving. The hotter the material, the faster the molecules are vibrating or moving around, which we can measure as a higher temperature. If two objects are in contact, heat energy flows from the hotter object (higher temperature) to the cooler object (lower temperature). In other words, at the point of contact between the two objects, the faster-moving molecules on one side transfer some of their motion to the slower molecules on the other side, until the two materials reach the same temperature.

Heat is not the same thing as temperature. Heat is a measure of how much energy is required to raise the temperature of a mass, which can vary considerably from one type of material to another. For example, it takes far more energy to heat water than to heat a chunk of gold or lead of the same weight. The concept of heat also takes into account the amount of material being heated; it takes more energy to heat a large object than a small one. The reverse is also true: if an object requires more energy to heat, it has more heat energy available to give off when it cools. The amount of energy absorbed or given up as the temperature changes is how we measure heat.

When you experience the warmth of sunlight or a fire in the fireplace, the heat you feel is not from direct contact with hot matter. An intermediate step involves light. Sunlight and firelight both include a lot of light energy in the infrared portion of the spectrum. Most types of matter, including your skin, are very good at absorbing photons in this range of wavelengths. Absorbing these photons causes your skin to heat up because as the photons disappear, the energy changes form from light energy to heat energy. Thus, radiant heat is not actually heat at all, but a form of light that turns into heat when it's absorbed.

As an illustration, a cup of coffee cools down for three reasons. First, both the coffee and the cup lose heat by giving off radiant energy (infrared light). Second, the coffee and the cup both transfer heat to the air through conduction. (Heat is also transferred to the surface the coffee cup sits on.) And third, the coffee loses heat energy due to evaporation (in the same way that sweating cools your body).

Chemical Energy

When most of us hear the word *chemical*, we think of dangerous materials with long, complicated names. We typically think of compounds that were synthesized in a laboratory or in a chemical factory. However, the scientific definition of *chemical* includes natural as well as synthetic compounds. Water is a chemical. Salt and sugar are chemicals. So is carbon dioxide. In fact, nearly every type of matter consists of chemicals.

To say that something is a chemical is essentially just saying it consists of molecules (especially if all the molecules are the same kind). A molecule is two or more atoms held together by atomic bonds, also called chemical bonds. One of the most interesting things about chemical bonds is that the creation of a bond either requires energy or gives it up, depending on the circumstances. Likewise, the breaking of a chemical bond either requires energy or gives it up, the opposite of what happened when the bond was created. The most fascinating cases are chemical bonds that cannot be broken unless additional energy is applied but that ultimately give off a lot of energy after the bond is broken. In fact, these kinds of bonds are fundamental to life.

In living creatures, eating food is all about acquiring molecules that are rich in chemical energy—fats, carbohydrates, and proteins. Most of this energy is stored in carbon bonds—that is, bonds that involve carbon atoms. In other words, each carbon atom in a molecule is bonded to other atoms, many of which are also carbon atoms. The result can be a relatively small and simple molecule, such as glucose, or a huge and complex molecule, such as a protein. We call such carbon-based chemicals organic compounds. Because organic compounds are full of chemical energy, this energy can be released as needed to drive

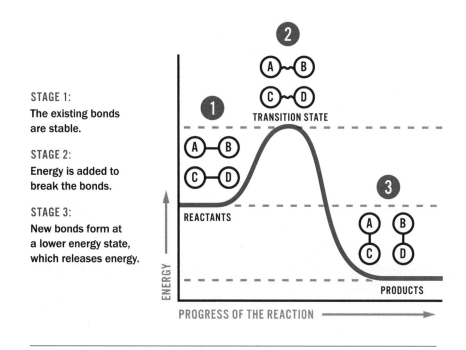

STAGE 1:
The existing bonds are stable.

STAGE 2:
Energy is added to break the bonds.

STAGE 3:
New bonds form at a lower energy state, which releases energy.

A generalized chemical reaction that gives off energy

essential biochemical processes, to provide energy to muscles, and to generate heat.

But where does all of this energy originally come from? It comes from green plants and cyanobacteria that conduct photosynthesis, which is the process of converting the energy of sunlight into chemical energy. Living creatures that cannot photosynthesize, such as animals, must acquire these energy-rich compounds from another source, usually by eating other living things. Thus, virtually all of the chemical energy in all of the plants and animals in the world can be traced back to sunlight.

Kinetic Energy

The word *kinetic* refers to something that exhibits motion, as in a kinetic sculpture. So kinetic energy is the energy of motion, or more precisely, the energy inherent in a mass due to the motion of that mass. Think of a moving car, a flying baseball, or a bullet in midflight. In each of these cases, the object possesses kinetic energy due to its motion.

As a kid, I was quite bothered by this concept. My reasoning was that if these objects actually possessed energy, it should be possible to examine each object to find evidence of the energy—the equivalent of photons for light, vibrating molecules for heat, or atomic bonds for chemical energy. But as far as I could tell, a moving object is physically identical to a stationary object. Furthermore, we were taught that all motion is relative and can only be measured in terms of a reference point (except for light, which always has the same speed, regardless of the reference point). Therefore, I reasoned, if my reference point is the moving object itself, that object has no motion and therefore no kinetic energy. Based on these internal contradictions, I figured I'd proven that kinetic energy does not actually exist. So why were my textbooks saying it does exist?

Many years later, I gradually came to understand the answers to my questions. The first answer is that you have to consider the entire system, not just the individual object. I really didn't like that answer, because it fails to define what the system is. The second answer is that energy only matters when it changes form. We need only consider how an object acquires—and gives up—its kinetic energy. When two objects crash into each other, it doesn't matter which one was moving or if both were moving. What matters is the speed of the two objects *relative to each other*. It's in that relative difference of velocity that kinetic energy plays a role when the two objects collide.

In a car, the energy to put the vehicle into motion comes from the chemical energy of the fuel. When we apply the brakes, the vehicle's kinetic energy is converted into heat energy in the brake pads. These pads are exposed to the air so that the acquired heat can quickly dissipate into the air. In a speeding bullet, the kinetic energy comes from the rapid release of chemical energy stored in the gunpowder. Although the mass of the bullet is small, its great speed means that it carries a lot of momentum, a concept closely related to kinetic energy. (Momentum equals mass times velocity.) On those rare occasions when an asteroid strikes the earth, producing a giant crater, the tremendous kinetic energy is due not just to the mass of the asteroid but also to its great speed relative to the earth.

Potential Energy

When I was a kid, the concept of potential energy seemed even less believable than the concept of kinetic energy. I developed several arguments that, in my mind, completely disproved the concept. Eventually I came to understand and accept the idea of potential energy, but I had to get past a few issues.

Potential energy exists in several forms, but let's start with gravitational potential energy—the type most often mentioned in textbooks. The idea is that if you lift something to a higher elevation, it gains potential energy due to the energy you expended to lift the object up. If that object then falls or rolls back down, the potential energy turns into kinetic energy, which means that the object is moving. As a result, an object that's falling (or rolling down a hill) tends to go faster and faster, as more of the potential energy changes into kinetic energy.

Consider a roller coaster. At the start of the ride, a machine embedded in the track slowly hauls a little train up a long slope to the highest point on the track. Then the train is released. Gravity pulls the train down each of the downhill slopes in the course, with the fastest speeds occurring at the bottom of the slopes. The train has enough momentum to rise up the next hill, which is not as high as the first hill. This continues until the train completes the circuit and is brought to a stop. The entire loop is an exercise in changing potential energy to kinetic energy and back again, over and over, until all of the energy has been lost due to friction. And it was all powered by the machinery that lifted the train up the first slope, which gave the train enough potential energy to complete the entire loop.

In contrast to a roller coaster ride, we have other uses of potential energy in which some of the resulting kinetic energy is captured and put to practical use. Some of the best examples involve hydropower, in which water rushing downhill gives up some of its kinetic energy to spin a large wheel or turbine, thereby generating electricity, or in older days, driving a mill. Mills were used to grind grain or drive machinery (as in a sawmill or textile mill) by direct mechanical coupling of the water wheel to the machinery. But if energy cannot be created or destroyed, where does the energy in hydropower come from? It comes from sunlight, which evaporates water into the air, which then rises into the sky (due to convection currents also caused by the sun), which later falls as rain and is collected in a river or reservoir.

It's tempting to say gravity provides the energy that powers a roller coaster or a hydroelectric plant, but that's incorrect. Gravity simply provides a force, allowing potential energy to be stored when a mass is moved against that force. The potential energy is then converted into kinetic energy by releasing the mass, allowing it to fall. Gravity plays a key role in the process, but gravity does not provide any energy itself. In both of our examples, the energy came from another source: the machinery in the roller coaster's first slope and the energy of sunlight.

This idea—working against a force to store potential energy—has many other applications. If you push the north pole of a magnet against the north pole of another magnet, you will feel a strong resistance. It takes energy to push them together, and as long as you hold them together the system contains potential energy. When you release those magnets, they will fly apart, as the potential energy is converted into kinetic energy. Another good example deals with the compression of gases. Unlike liquids or solids, gases can be compressed into dramatically smaller volumes, but the more you compress a gas, the more pressure it has. Compressing a gas stores potential energy, and releasing the gas usually results in kinetic energy, as in an aerosol spray can.

A similar example involves stretching a rubber band, which puts potential energy into it that is converted into kinetic energy when you release it. If the rubber band is a slingshot, the kinetic energy can be transferred to a stone or other small object. Likewise, a spring that's stretched or compressed acquires potential energy, which is converted into kinetic energy when it springs back to its original shape. These examples of potential energy are sometimes called elastic energy. A particularly interesting example is a bow and arrow. When you pull on the bowstring, the bow is temporarily deformed, storing potential energy. When the bowstring is released, the bow snaps back to its original shape, transferring part of the energy to the arrow as kinetic energy.

So despite my initial reluctance to acknowledge the existence of potential energy, I now see that the concept is quite valuable. But instead of seeing potential energy as something that resides in an object, comparable to heat energy, I see it as the potential stored when a mass is moved against the resistance of a force, which could be gravity, air pressure, elasticity, magnetism, or whatever. In each of these cases, when the mass is later released, the force converts the potential energy to kinetic energy.

Electrical Energy

As a kid reading my science textbooks, I believed electrical energy made more sense than either kinetic energy or potential energy because I could point to a clear physical cause—the movement of electrons. Based on my interpretation of what I read, I developed a slightly flawed mental model that did a reasonable job of explaining electricity, but only in the case of direct current (as in a flashlight or battery-powered toy). My mental model failed to explain alternating current and static electricity, so those two concepts never made any sense to me.

To get a better understanding of electricity, I had to change my mental model. Contrary to my earlier belief, electrical wires are not filled with billions of loose electrons, flowing like water from the power station to my house. The electrons actually belong to the atoms of copper (or some other metal) in the wiring. Each atom of copper has twenty-nine protons and twenty-nine electrons. Normally, the twenty-nine electrons remain in orbit around the corresponding atomic nucleus, but every now and then one wanders off, briefly leaving the vicinity of its home atom. In the meanwhile, another wandering electron might jump into the available opening. No problem—the wandering electron easily finds another opening in a nearby atom because some other wandering electron has left a vacancy. The result is like a chaotic game of musical chairs, with electrons constantly hopping between atoms. This attribute of wandering electrons is what makes copper such a good conductor of electricity.

Under normal circumstances, all of this hopping around does not produce electricity, because the directions of movement are random and cancel each other out. But if a voltage is applied to the wire, it forces all of the wandering electrons to head in the same direction, which puts pressure on the wandering electrons farther down the line. This electrical pressure can be tapped to run motors or generate heat. The individual electrons in the system don't need to make the entire journey from the power station to my house; all they have to do is to put pressure on the neighboring electrons in the wires.

In the case of alternating current (AC), the electrons are forced to reverse their direction of travel many times each second. However, there is electrical pressure regardless of which direction the electrons are being pushed; either direction around the circuit would eventually get to my house. The pressure briefly disappears each time the current changes direction, but otherwise the pressure is continuous. Because of this brief loss of electrical pressure with each change of direction, electric lights running on AC actually flicker, but usually too fast for our eyes to detect.

For many of us, the most familiar example of static electricity is when we walk across a carpet and then get shocked when we touch a metal door handle. The other obvious example, on a very different scale, is a lightning storm. In both of these cases, a surplus of electrons builds up in a material, leaving a deficit of electrons somewhere else, but the electrons can't flow away because any potential path is blocked by an electrical insulator (such as air). Electrical pressure builds up, but for a while nothing happens. Eventually, something happens that allows the electrons to flow away in a single dramatic burst.

Other Forms of Energy

Several additional forms of energy are worth mentioning, including waves in matter, nuclear energy, and mass energy.

WAVES IN MATTER

Unlike light energy, which can travel through the vacuum of space, other kinds of waves exist only in matter. Sound waves, which are compression waves traveling through the air, provide the most familiar example. (Such waves are frequently called longitudinal waves, because the forward-and-back vibrational movement occurs in the same direction the wave travels.) A sound is generated when a moving or vibrating object gives a shove to the surrounding air molecules, which in turn shove the neighboring air molecules. As each molecule shoves its neighbors, it passes along its kinetic energy, producing a wave that travels through the air at more than 700 miles per hour. This may seem fast, but it amounts to only about 1,000 feet (1/5 mile) per second, so it takes five full seconds for sound to travel a single mile (three seconds to travel a kilometer). Based on this knowledge, you can estimate how far away a lightning strike is by counting the seconds between the flash and the boom.

Sound energy, in the form of compression waves, can also pass through liquids and solids. You can easily hear sounds when you are underwater, such as the noise from motorboats. But in the case of solids, a second kind of wave can also exist: shear waves, which occur when molecules vibrate from side to side perpendicular to the direction of the wave, rather than forward and back as in compression waves. Shear waves are frequently called transverse waves because of this perpendicular vibration.

Waves that occur on the surface of oceans and large lakes, at the interface between water and air, are another example of waves in matter. Such waves are quite different from sound waves in water, yet in both cases the molecules of water pass along their kinetic energy while hardly moving from their original location.

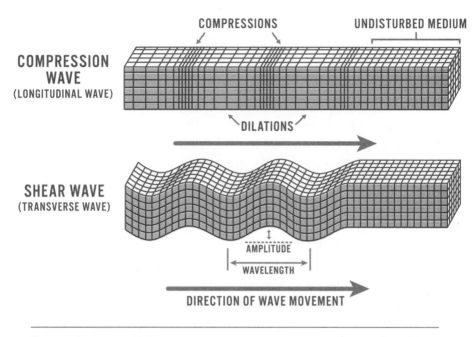

COMPRESSION WAVE (LONGITUDINAL WAVE)

COMPRESSIONS

UNDISTURBED MEDIUM

DILATIONS

SHEAR WAVE (TRANSVERSE WAVE)

AMPLITUDE

WAVELENGTH

DIRECTION OF WAVE MOVEMENT

Compression waves and shear waves

Earthquakes also involve waves passing through matter, but this topic is a bit more complex. A typical earthquake generates several kinds of waves, consisting of body waves (which can travel deep underground) followed by surface waves (which only occur on the surface of the earth). The body waves are subdivided into primary waves (which are compression waves) and secondary waves (which are shear waves), while the surface waves are divided into Love waves and Rayleigh waves. Love waves (named after British geophysicist Augustus Edward Hough Love) shake the ground back and forth, while Rayleigh waves (named after British scientist John William Strutt, Third Baron Rayleigh) are similar to ocean waves, resulting in an up-and-down motion.

NUCLEAR ENERGY

Nuclear energy can come from either of two distinct sources: the fission (splitting) of the nuclei of very large atoms or the fusion (merging) of very small atoms. The energy of a nuclear power station comes from fission, while the energy of the sun comes from fusion. The earliest atomic bombs (in the 1940s) relied on the fission of uranium or pluto-nium, but by the 1950s scientists had learned to create even stronger bombs based on the fusion of hydrogen.

Nuclear energy exists because at the scale of atomic nuclei, two very important forces tug in opposite directions. One force is the elec-trostatic force (also called the Coulomb force), and the other is the nuclear force. The electrostatic force causes atomic particles with opposite charges (such as protons and electrons) to attract each other, and particles with similar charges to repel each other. The nuclear force causes protons and neutrons to be attracted to other protons and neutrons, but only when they are very close to each other, no farther apart than the diameter of a proton. The nuclear force is why protons can clump together into a nucleus despite having the same charge.

The upshot is that when you split a very large nucleus, you first have to overcome the nuclear force, but once you do, a huge amount of potential energy is released due to the electrostatic force. When you merge two very tiny nuclei, you first have to overcome the elec-trostatic force, but once you do, a huge amount of potential energy is released due to the nuclear force.

MASS ENERGY

While nuclear energy can be explained in terms of potential energy, due to the tension between the nuclear force and the electrostatic force, a very different explanation says that the energy of a nuclear bomb (or a nuclear power plant) comes from converting matter into energy. So which is it? In fact, both explanations are correct.

After an atom of uranium splits into two atoms—or after two atoms of hydrogen fuse to form an atom of helium—you can count up the total number of protons and neutrons (including any neutrons that were ejected or absorbed), and you've got the same number as before. So how can we say that some of the mass was converted into energy? The answer is that the total mass of the end products is slightly less than the mass of the original materials, despite the fact that no nucle-ons (protons and neutrons) were destroyed. To make sense of this, you must understand that the mass of a nucleus does not exactly equal the total mass of its nucleons; this is because the mass is also affected by the binding energy that holds the nucleus together. Consequently, the total mass tends to change when nuclei split or fuse.

This idea may seem completely counterintuitive, but it is consistent with Einstein's famous equation: $E = mc^2$. This equation has become a meme, familiar to almost everyone even if we don't understand it. The E stands for energy, the m for mass, and the c for the speed of light (which never changes). Because E sits on one side of the equation and m on the other side, this equation says that either one can be converted into the other. In fact, this equation says that energy and mass are equivalent—essentially the same thing, just in a different form. In other words, you could say that mass is simply another form of energy. This explanation sounds way too freaky, but it's a helpful way of understanding not only nuclear energy but also some of the relativistic effects that occur when objects travel at nearly the speed of light.

Energy and the Human Body

We have now reviewed a fairly complete list of the various forms energy can take. Many of these forms of energy have roles in human biology, but not all to the same degree.

ENERGY WITHIN YOUR BODY

Chemical energy—the energy you get from food—plays one of the biggest roles in human biology of any form of energy. Fats, carbohydrates (sugars and starches), and proteins are all energy-rich foods, although it is primarily the fats and carbohydrates from which you extract the chemical energy. (Protein molecules, although rich in energy, are too valuable to burn as fuel, except in an emergency.) If you overstock your body with too much chemical energy, the excess is stored as fat.

Your body uses chemical energy for several purposes. One purpose is to contract your muscles, including the internal muscles that drive the heart and lungs. The contraction of your muscles converts chemical energy into kinetic energy. Muscles allow you to walk and run, pick up objects, and make all sorts of other movements. But a lot of your chemical energy is used for another purpose: to drive the biochemical processes that keep your cells alive and your body operating. In particular, your body constantly manufactures crucial organic compounds, reconfiguring the atoms obtained from the food you eat. These compounds are used for many different purposes in cells throughout your body. Thousands of different enzymes in your body mediate these processes, most of which require the expenditure of chemical energy.

Remember that energy cannot be destroyed, only converted to a different form. Your body's use of chemical energy eventually converts much of that energy into heat. That is why vigorous exercise causes your body to warm up so dramatically. But burning up your food energy is not your body's only source of heat. You constantly exchange heat energy with everything around you, through direct contact (heat conduction) and by exchanging infrared radiation (radiant heat). Infrared radiation is actually a form of light, which means that light helps to keep you warm, in addition to allowing you to see.

Electrical energy also plays an important role in the human body, allowing messages to travel along your nerves and controlling the rhythm of the heart. However, this concept is easily misinterpreted. Nothing in your body conducts electricity as well as copper wire, so electrical energy in your body does not behave like an electric circuit. Instead, specialized cells called neurons, located in your brain and nervous system, use an electrochemical system to communicate with neighboring neurons. In other words, each neuron uses a process that is partly electrical and partly chemical to communicate with its neighbors. The odd thing about this system is that electrical signals are used *within* the neuron, while chemical signals are used to bridge the gap *between* neurons.

Electrodes attached to the skin are capable of detecting the electrical activities of neurons within the body. An EKG (electrocardiograph) measures the firing of neurons that control the timing of the heart. An EEG (electroencephalograph) measures the firing of neurons in the brain.

Because neurons in the brain tend to fire in a coordinated fashion, an EEG will detect "waves" of electrical activity—in other words, a pattern of oscillating increases and decreases of activity. Different rates of oscillation correspond to various levels of alertness or sleep. These patterns are called neural oscillations, but the popular term is *brain waves*. Unfortunately, this latter term is highly misleading. It seems to imply a unique form of energy, but brain waves are simply patterns of electrical activity, similar to the heartbeat pattern generated by neurons in the heart that an EKG picks up. (Notice that we don't call EKG patterns *heart waves*.)

Sound energy doesn't play much of a role within the human body, but it's certainly important in terms of your ability to hear what goes on around you and to communicate through speech.

ENERGY YOUR BODY GIVES OFF

Your body can transfer energy to the world around you in several different ways:

- **HEAT ENERGY** If any part of your body is in direct contact with a colder object, heat energy moves from your body into that object, cooling your body and warming that object. Of course, some materials are much better than other materials at wicking away this energy. Water is very good at conducting heat, while air is a poor conductor of heat. In addition to conduction, evaporation is another factor. If your skin is wet with water or sweat, evaporation can transfer a lot of heat from your skin into the air.

- **RADIANT ENERGY** Every object, including your body, gives off infrared radiation (IR), a form of light not visible to our eyes. Warmer objects give off far more IR than cool objects, but even cold objects give off some IR. The wavelength of this light depends on the temperature of the object, shifting to shorter wavelengths for hot objects. Some types of night-vision goggles work by converting IR light to visible light. That is why warm objects such as humans stand out so sharply in such goggles.

- **KINETIC ENERGY AND POTENTIAL ENERGY** Whenever you move an object, you transfer kinetic energy to that object. If you're acting against a force, such as lifting an object against the force of gravity, you're providing that object with potential energy.

- **SOUND ENERGY** Every time you speak or sing, or clap your hands, or make any other kind of noise, you're converting kinetic energy into sound energy. The kinetic energy originates from your muscles, which in turn are powered by chemical energy.

- **CHEMICAL ENERGY** Whenever any energy-rich molecules escape from your body, you could say you've lost chemical energy to the world around you. The main source of these lost organic molecules is your feces, for the simple reason that your body is not capable of digesting all the materials in your food. In particular, humans are not able to digest cellulose, which has just as much energy as sugars and starches. Urine and sweat also contain small amounts of energy-rich organic compounds. Any gas you pass typically contains organic compounds too.

- **ELECTRICAL ENERGY** Although electric eels can give off jolts of electrical energy, humans cannot. (Eels have special organs for the purpose, and they are surrounded by water instead of air.) Thus, we humans don't leak much electrical energy into the world around us. A small exception is when you build up a static charge and then discharge it by touching something, such as a door handle or a cat.

That essentially covers the entire list of forms of energy that are important to the human body: chemical, kinetic, heat, light, electrical, and sound energy. Yet if you spend much time on the internet looking for information about energy and the human body, you're likely to encounter references to forms of energy not recognized by science, such as psychic energy, spiritual energy, life energy, healing energy, thought energy, mind energy, crystal energy, and negative energy. Likewise, most references to vibrational energy, vortexes, auras, and "the body's energy field" also deal in information not recognized by science. Why does science refuse to recognize the validity of these topics? Science is based on the development and refinement of testable concepts. Some of these "new age" concepts are framed in a manner that's not testable, and others fail to produce the predicted results when examined under rigorous scientific testing.

Giving Up on a Better Definition of *Energy*

After all of this talk about energy, I still have not provided a definition of the word, a nice little phrase that applies equally well to all different forms of energy. The definition I was taught in high school, "the capacity to do work," just passes the buck because now we need to define *work*. I was told that work is "moving a mass against a force," which covers only a subset of what energy can do. For example, it ignores the energy expended when heating an object. So a better definition of *energy* is "the capacity to heat a mass or to move a mass against a force." But I still don't like this definition because the word *capacity* is so vague.

Perhaps I should say that energy is "a physical quantity that is transferred when a mass is heated or moved against a force." But the phrase "a physical quantity" is still terribly vague. Worse, this definition doesn't seem to acknowledge photosynthesis or most other biological processes that involve chemical energy. My attempts to define *energy* continue down this endless spiral, never reaching a satisfactory conclusion. It's like the day I stopped at a country store to ask for directions and was told, "Sorry, sir, but you can't get there from here."

Other than the vagueness, what really bothers me about all these definitions is that none of them does a decent job of priming my intuition. If I were hearing about energy for the first time and the first step was to read the definition, I would be left scratching my head. I'd see very little connection to the topics of light, heat, kinetic, potential, electrical, wave, nuclear, and mass energy. In fact, the definition of *energy* ignores nearly everything that's interesting about the various forms of energy. For me, the essential big-picture concept I've learned about energy is not its definition but the fact that energy cannot be destroyed; it can only change form. Furthermore, most of the interesting things about energy appear only when energy changes form.

There is so much more I could say about energy. There is so much more I feel I *ought* to say. I could write an entire book on the topic and still not get to everything I would hope to cover. On the other hand, writing this chapter has worn me out, and now I'm completely out of energy. So I think I'll just stop here.

11

Left-Brained and Right-Brained

Anyone who knows me well has noticed that I tend to be logical, analytical, and methodical, sometimes to the point of quirkiness. From my own point of view, my behavior makes perfect sense no matter how odd it might seem. For example, every time I do laundry, after carefully folding everything in a very specific manner, I place all the socks, underwear, handkerchiefs, and polo shirts in my dresser according to a strict protocol. Each of these categories of clothing is organized into its own queue. Each newly washed item enters the back of the appropriate queue, and the items I wear each day are drawn from the front of the queues. As a result of this policy, every item in each category receives an equal amount of use.

And then there is my extreme reluctance to cut corners, both literally and figuratively. If the sidewalk makes a right-angle turn, by golly I'm going to make a right-angle turn rather than take some silly short-cut that would force me to leave the sidewalk.

My family and friends are accustomed to my habits and usually find my behavior humorous rather than annoying, but they often can't resist smiling and making a comment. On many occasions I have been told that I am obviously a left-brained person. I've found this to be an interesting observation, but it always made me wonder: What are the actual details of this concept of left-brained versus right-brained people and to what extent is this model supported by science?

The Concept of Left-Brained and Right-Brained People

The popular idea that people are either left-brained or right-brained is analogous to the well-known fact that people tend to be either left-handed or right-handed. If you are right-handed, as the majority of people are, your right hand is said to be dominant. You prefer to use your right hand instead of your left on nearly any task that requires only one hand, or on a task that requires that one hand take a bigger role than the other. The concept of being either left-brained or right-brained holds that a similar effect is true for the brain—that one side of the brain is almost always dominant over the other—so you will use that one side of the brain a great deal more than the other side.

The second key part of this idea is that different kinds of mental tasks are allocated to one side of the brain or the other. In this model, the left side of the brain is used for tasks that are analytical or methodical. The right side of the brain is used for tasks that are creative or artistic. According to this explanation, your personality is strongly influenced by which half of your brain is dominant. In other words, a left-brained person is said to be logical, analytical, and objective, while a right-brained person is intuitive, imaginative, and artistic. However, as we will see shortly, this model is seriously flawed.

The idea that people might be left-brained or right-brained first became popular in the early 1970s, loosely based on results from brain research that had begun the previous decade. The largest part of the brain—the cerebrum—is in fact separated into two distinct halves. The left half and the right half communicate with each other through a bundle of nerve fibers called the corpus callosum, but otherwise the two halves are separate organs. In terms of appearance, these two organs look like mirror images of each other—quite symmetric.

If a person suffers from debilitating epileptic seizures, and if those seizures cannot be controlled by medicine, one possible treatment is to surgically sever the corpus callosum. Oddly, this dramatic procedure is usually quite effective, and the people who receive the treatment are usually able to resume normal lives. However, as brain researchers began to study these people, they soon noticed some fascinating side effects resulting from a lack of communication between the two halves of the brain.

Before the beginning of this research in the 1960s, it was already known that the left side of the brain controls the muscles on the right side of the body, while the right side of the brain controls the muscles on the left side of the body. This fact had been deduced by studying people who had suffered an injury to one side of the brain with subsequent paralysis on the opposite side of the body. By studying people whose corpus callosum had been severed, scientists hoped to gain additional insights into how various mental tasks are divided between the two halves of the brain.

Scientists devised a great number of experiments using people with a "split brain," but the general idea was to expose only one half of the person's brain to a piece of information and then to explore the consequences. In one such test, the subject was blindfolded and an object was placed in either the left hand or the right hand. The object was in the shape of a numeric digit—1, 2, 3, or 4. The person was asked to feel the object in order to identify the number. Next, the person was asked to verify the identification by holding up a matching number of fingers using the same hand that had felt the object. Finally, the person was asked to say out loud what the number was.

If the person felt the object with her right hand, everything went smoothly. She would hold up the correct number of fingers and then say the correct number out loud. But if the person felt the object with her left hand, the results would be different. She would hold up the correct number of fingers, but then she would usually say the wrong number. It soon became clear that she was simply guessing the value of the number when she spoke, even though she had already signaled the correct answer with her left hand.

This and other similar experiments revealed several fascinating details about the brain. For example, they showed that the ability to speak is localized in the left side of the brain, the same side of the brain that communicates with the right hand. And they showed conclusively that the left and right sides of the brain do not function as exact mirror images. Instead, certain functions of the brain are focused on one side or the other, in an asymmetric manner.

The discovery of asymmetry in the functioning of the brain fueled a lot of speculation as to the extent and nature of the asymmetry. Scientists and the general public alike were fascinated with the possibilities. While scientists began the process of investigating this phenomenon in more detail, the idea took on a life of its own, magnified and nurtured by the popular media. In the early 1970s, several well-known national magazines published articles stating that people tend to be either left-brained or right-brained, and that this influences your personality. The concept soon became widespread, promoted by an avalanche of pop-psychology articles about how to use this knowledge to understand other people or to optimize your own functioning.

Problems with the Left-Brained/ Right-Brained Concept

The concept of left-brained and right-brained people is certainly a fascinating model, and because of its popularity, a lot of people assume it's true. But what does science have to say about it? Although this model seemed reasonable when it first became fashionable, later research showed it to be untrue. The concept has two principal problems:

- **PROBLEM 1:** *There is no dominant side of the brain.* Subsequent research used fMRI (functional MRI) to map the brain activity of thousands of people as they did different tasks or just rested quietly, doing nothing at all. In each brain scan, locations where a lot of energy was being used would light up. When a person switched tasks, a different map of mental activity would appear in the scan. Extensive testing of this sort showed that neither side of the brain is more active than the other side, regardless of your personality type or how you tend to think. It turns out that all of us use both sides of the brain in relatively equal amounts and that the two sides work together simultaneously. Because the brain has no dominant side, it's a myth that lateral dominance determines your personality and way of thinking.

- **PROBLEM 2:** *Thinking is far less lateralized than the popular model suggests.* The second issue is a bit more subtle, because science has indeed shown some lateralization in the brain. In other words, certain specialized subtasks are handled primarily by either the left brain or the right brain. This doesn't mean that when you tackle a logical problem, only the left side of the brain lights up, or that when you tackle a creative problem, only the right side of the brain lights up. Regardless of the task, your brain assigns subtasks to various areas on both the left and right sides of the brain. But certain subtasks, such as speaking your thoughts out loud, might be handled primarily by just one side. This doesn't make you left-brained or right-brained, but it does shed light on how the brain works.

The upshot is that the model of left-brained versus right-brained people was disproven long ago, even though in popular culture the idea never completely died, but instead became an enduring myth. That said, it is true that different regions of your brain play different roles in your thinking and that these differences are very important. But if these distinct regions are not the left brain and the right brain, what are they?

Specialized Regions in the Cerebrum

The human brain consists of three main parts, but if you look at it from the top, all you can see is the part called the cerebrum. You can clearly see two distinct halves or hemispheres—in fact, the corpus callosum (the connector between the two halves) is not even visible. But if you look at the brain from the side or rear, you can see the cerebellum and the brain stem beneath the cerebrum. In addition to these three parts, the brain has several smaller parts that are also quite important. (We'll come back to those later.)

Each hemisphere of the cerebrum is divided into four lobes: frontal, temporal, parietal, and occipital (see illustration on page 208). So there's a left and a right frontal lobe, a left and a right temporal lobe, and so on. Different types of brain activity tend to be focused in specific lobes:

- frontal lobe: movement of the body (motor control), speech, smell, emotional regulation, concentration, planning, and problem solving
- temporal lobe: hearing, facial recognition, long-term memory, and language comprehension
- parietal lobe: touch and pressure, taste, and body awareness
- occipital lobe: vision

Because the four brain lobes have such distinct specialties, it's surprising the popular media has never promoted a model in which people's personalities are tied to the dominance of a particular lobe rather than a particular hemisphere. I can imagine someone saying, "You're so frontal lobed!" or "You're clearly a temporal-lobed person." Although such a model would be no more realistic than the model of left-brained and right-brained people, at least it would support four types of personalities instead of just two. But in actuality, associating a particular type of brain activity with an entire lobe is rather imprecise. In most cases, the different types of brain activity tend to be focused in specific parts of the corresponding lobe rather than spread across the entire lobe.

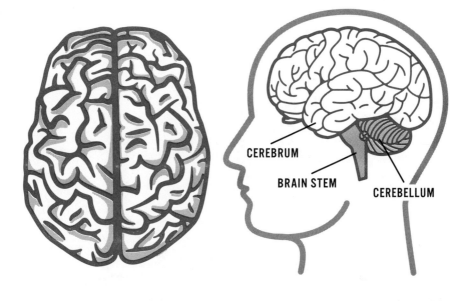

CEREBRUM

BRAIN STEM

CEREBELLUM

Top and side views of the human brain

For most of these specialties, both sides of the brain participate. As an example, both the left and right sides of the brain have zones specialized for hearing, taste, and smell, and in most cases the two sides participate in roughly equal measure. The biggest exception is speech generation, which in most people is highly concentrated in the left hemisphere of the brain, in the Broca and Wernicke areas (although the thought processes related to speech involve other parts of the brain as well). But even this generality is not universally true. In about 95 percent of right-handed people, speech is indeed concentrated in the left hemisphere, but only about 50 percent of left-handed people exhibit this phenomenon. In other people, speech is either concentrated on the right side of the brain or divided fairly evenly between the two sides.

This illustrates the fact that the brain is rather plastic—that is, the zones of specialization can vary somewhat from person to person, for a variety of reasons, including adaptation to a particular environment (especially during childhood). This plasticity is especially helpful when brain damage occurs, as it allows other parts of the brain to develop skills to help fill in for the damaged areas. Brain plasticity is

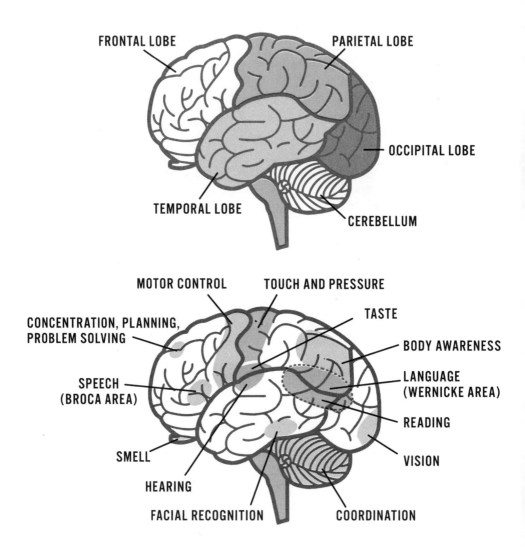

Lobes and functional zones in the brain

also evident in people who were born either blind or deaf; the brain develops in a different manner in response to the different balance of sensory stimulation.

Other than speech, the most noteworthy example of asymmetric lateral specialization in the brain is facial recognition, which tends to be concentrated on the right side of the brain.

Other Parts of the Brain

After the cerebrum, the largest and most visible part of the brain is the cerebellum. Personally, I think the words *cerebrum* and *cerebellum* are too similar, and I sometimes get them confused. To remind myself which is which, I say, "The 'bellum' is *below* and *behind*, while the 'brum' is *up* by the *brow*." It's silly but effective, although not all of the cerebrum is up by the brow; only the frontal lobe is.

Because the cerebellum is snuggled up beneath the cerebrum, you might assume (as I once did) that the two have a direct connection. That turns out not to be the case. Instead, they're both connected to the brain stem, which in turn connects the brain to the spinal cord. The cerebellum plays a major role in coordination, ensuring that the various parts of the brain work together in a smooth manner. The net result reflects the other meaning of the word *coordination*—that the muscles of your body work together in a coordinated fashion. The brain stem is composed of three parts: the medulla, the pons, and the midbrain. The brain stem controls many of the body's vital involuntary functions, such as heart rate, breathing, blood pressure, and digestion.

Hiding underneath the cerebrum are several small but important parts of the brain, including the thalamus, the hypothalamus, the hippocampus, and the pituitary gland. You've probably heard of all of these little parts without necessarily knowing where they're located, much less what they do. For me, the words *thalamus*, *hypothalamus*, and *hippocampus* were always a blur—three easily confused organs performing easily confused tasks. But they really do have distinct roles:

- The thalamus is a kind of switching station that routes signals between the cerebrum and the rest of the brain and nervous system. Such signals include sensory information coming from elsewhere in the body, as well as outgoing messages to the muscles of the body. The thalamus is also involved in sleep and consciousness.
- The hypothalamus is a kind of thermostat, regulating your body temperature. The hypothalamus is also involved in emotions, hunger, thirst, and circadian rhythms.
- The hippocampus is responsible for creating memories. A related function is to help you navigate your environment based on what you remember about it.
- The pituitary gland is a tiny part of the brain that controls various hormones, sending chemical signals to other parts of the body. It's amazing how many roles this tiny gland performs (either alone or in conjunction with other parts of the brain), affecting your growth rate, metabolism, digestion, breathing, blood circulation, and levels of sugar and water in your body.

These aren't the only little parts hiding in this zone of the brain. Other examples include the pineal gland (another little gland that releases hormones) and the amygdala (which plays an important role in processing emotions).

Notice that the illustration showing the locations of these little organs depicts the right side of the brain viewed from the left side, as though the entire left side of the brain were cut away. If you examine the picture carefully, you can see where the brain is naturally divided in two versus the parts that you would have to cut to reveal this view. You can see sliced cross sections of the corpus callosum (which connects the two halves of the cerebrum), the brain stem, and the cerebellum. In other words, these parts of the brain are fully connected between left and right, in contrast to the two separate hemispheres of the cerebrum.

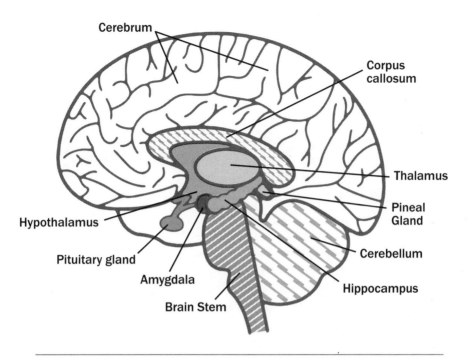

Front-to-back cross section of the brain

Neurons and How They Work

Now that we've identified the various parts of the brain and the roles they serve, the next step is to take a peek at how the brain works at the cellular level. To put it another way, the brain is made of living tissue and therefore is composed of living cells. But what exactly are those cells, and what functions do those cells perform?

The cells of the brain and the nervous system can be divided into two categories: neurons and glial cells. The role of neurons is to relay messages using direct cell-to-cell communication. Some of these messages go from one part of the brain to another. Other messages travel to or from the brain, connecting the brain to the rest of the body. Glial cells of several different kinds provide various forms of support and assistance to the neurons. Some glial cells form the myelin sheaths that protect parts of the neurons. Other glial cells provide metabolic support, while still others form membranes to divide one area from another.

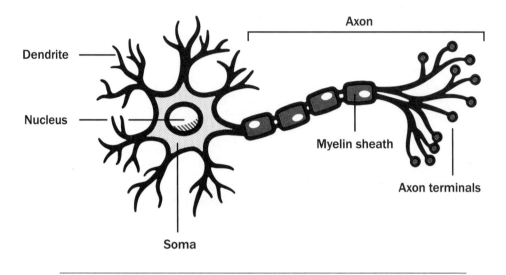

Anatomy of a neuron, showing the distinctive shape

The adult human brain contains roughly eighty billion neurons, and there are additional neurons in the spinal cord and elsewhere in the nervous system. Most neurons have a distinctive shape that facilitates their communication function. A typical neuron has a long, narrow extension—called an axon—that protrudes from the cell body (soma) like a thread. An axon can be up to 3 feet long, although most axons are much shorter. This is how a neuron can have direct connections to other neurons in distant parts of the body. Because of branches on the axon, a single neuron can often send messages to many other neurons. Elsewhere on the same neuron, other branches called dendrites are capable of receiving messages from the axons of other neurons. Consequently, a single neuron might be connected to hundreds of other neurons, including neurons on the opposite side of the brain.

One of the defining characteristics of a neuron is that its communication system is electrically powered, although in a manner quite different from how you might normally think of electricity. Each neuron uses chemical energy to maintain a voltage gradient across its cell membrane. It does this by selectively pumping positively and

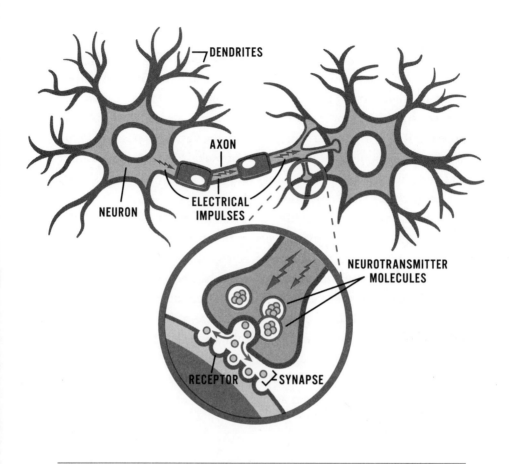

Connections between neurons and the details of a synapse

negatively charged ions (including sodium, potassium, chloride, and calcium) across the cell membrane. The result is a difference in voltage on either side of the membrane—that is, inside and outside the cell. This acts something like a battery, storing a charge. When a neuron fires (as a result of input from other neurons), it suddenly allows many of those ions to flow back through, releasing the charge. The resulting change in the internal voltage of the cell immediately flows down the axon of the neuron to the various axon terminals that are in contact with other neurons.

The connections between neurons are called synapses, and each connection includes a tiny gap between the two cells. Communication across each synapse is accomplished via chemicals called neurotransmitters, which travel across the gap to receptors on the other side. The human brain has trillions of synapses—a mind-boggling quantity—but they don't all work in the same manner. Some synapses excite the neighboring neuron, some inhibit the activity of the neighboring neuron, and some produce more complicated interactions. The huge number of synapses, combined with the variety of possible interactions between neurons, is part of what makes the brain so complex and powerful. For each neuron, the combined effect of all of this input determines when and how often the neuron fires.

Note that when a neuron fires, the electrical signal never leaves the neuron. The electrical signal travels up the axon to all of its outgoing synapses, but the communication across the synapses is accomplished via the neurotransmitter chemicals. Because the firing of a neuron is accomplished by allowing the sudden movement of charged ions across the cell membrane, it does have a detectable effect on the voltage of the fluid surrounding the neuron. These small voltage fluctuations travel through the brain fluid and can be detected by electrodes placed on your scalp. This is what allows an electroencephalogram (EEG) to record brain activity.

Gray Matter and the Cerebral Cortex

Now that we have taken a peek at how neurons work, we can zoom out to see how the arrangement of neurons affects the appearance and function of different regions of the brain.

Earlier I mentioned that the corpus callosum, which connects the left and right halves of the cerebrum, is composed of nerve fibers. Now we can make more sense of that statement. Each of these nerve fibers is actually the axon of a neuron. In some cases the body of the neuron is in the left side of the brain, which means that the axon crosses over to send messages to the right side of the brain. In certain other cases the body of the neuron is on the right side, sending messages over to the left side.

The brain and spinal column are composed of regions of gray matter and regions of white matter. The corpus callosum is an example of white matter; it actually looks white due to the white color of the myelin sheaths wrapped around the axons. If you view a cross section of some other part of the cerebrum, you'll see gray matter appearing as a thick layer near the surface of the brain, while the inner regions of the cerebrum are composed of white matter. The gray matter consists primarily of the bodies (soma) of neurons (which lack the myelin sheaths), while the white matter is composed primarily of axons, just as in the corpus callosum.

This arrangement, with the soma on the outside of the brain and the long-distance axons running through the interior, is an efficient arrangement for communication among the neurons. I think of the white matter as a highway system connecting all those suburban cul-de-sacs in the gray matter. However, in order to squeeze more neurons into the cranium while still maintaining this arrangement, the brain must increase its surface area as an embryo develops into a young human. The growing surface area results in the characteristic deep folds in the cerebrum. The deepest folds separate each half of the cerebrum into four distinct parts, the four lobes already discussed.

The term *cortex* refers to the outer layer of certain organs in the body. The gray matter of the cerebrum is typically called the cerebral cortex. Likewise, the gray matter of the cerebellum is called the cerebellar cortex. The region of the brain that enables vision—located mostly in the occipital lobe of the cerebrum—is called the visual cortex.

The Brain's Activities

The various activities of the brain can be categorized in several possible ways. One important distinction is between processes that are controlled by the conscious brain and those that are mostly or completely automatic. We typically mention the heart and lungs when we discuss the autonomic processes, but in fact nearly all of your internal organs respond to messages from your brain without any conscious thoughts on your part. For example, when your body detects that a meal has arrived in your stomach, your brain sends a message to the gall bladder to release bile into the digestive system.

Perhaps because of my background in computer science, my preference is to divide the brain's activities into four groups: input from the senses, output to the muscles, internal processing within the brain, and storage and retrieval of memories. I'll briefly discuss two of those categories, starting with input from the senses.

SENSORY INPUT

Although we like to say that humans have exactly five senses, a strong case can be made that we actually have more than five, as explored in an earlier chapter. What all senses have in common is that specialized nerve endings somewhere in the body are capable of detecting certain phenomena. When these nerve endings are triggered, messages are sent to the brain. The brain then has to make sense—in two different meanings of the word—out of these raw messages.

Much of this sensory input is triggered by phenomena that originate outside of the body, but important messages are also triggered by phenomena that occur entirely within the body. For example, we can sense when we have eaten too much because nerve endings in the stomach detect an excessive amount of stretching, and these nerve endings send messages to the brain. The five traditional senses (sight, hearing, smell, touch, and taste) are all based primarily on the detection of external phenomena—light, sound, airborne molecules, contact with external objects, and molecules in food and beverages. This means that our sense organs must have specialized structures that can somehow detect these phenomena and then generate nerve signals that communicate significant information about the phenomena.

An especially intriguing example is our sense of sight. To put it succinctly, your eyes detect light, but it is your brain that produces vision. The light receptors are located in the retina at the back of each eye, consisting of rods (which detect a broad spectrum of light) and three kinds of cones (each of which specializes in a narrow spectrum of light). Biochemical processes in the rods and cones convert "I detect something" or "I don't detect anything" into messages that are sent along the optic nerve to the visual cortex in the back of the head. That's when the real magic occurs, as the brain converts these signals into a dynamic stereo image of the world in front of you.

It's hard to appreciate the immense amount of data processing that the brain performs so that you can have vision. Vision consists of a whole lot more than simply passing along input from the eyes. In fact, a huge part of the human brain—the entire occipital lobe plus other bits—is dedicated to visual processing. For example, it is the brain—not the eye—that recognizes faces, identifies the edges of shapes, and notices movement. One particularly amazing aspect of how the brain processes data sent from the eyes is that the brain typically deals with any missing data by inventing plausible new data to complete the picture. The net result of all these processes is that what the brain perceives is often very different from what the eyes actually see.

The other sense I find especially interesting from the standpoint of sensory input to the brain is smell. Thousands of specialized nerve endings in the nose are capable of detecting a wide range of airborne molecules. The resulting messages are sent to the brain via the olfactory nerve. So far, this story sounds a lot like the story of vision, except that the nerve endings respond to airborne molecules instead of light. However, there are several major differences:

- In vision, only four distinct types of receptors send messages to the brain. Vision is possible due to the brain knowing the precise *location* of each receptor, which allows the brain to assemble a two-dimensional map for each eye. (Comparing the maps from the left eye and the right eye is part of how the brain constructs a perception of three dimensions.) In contrast, the nose contains about four hundred distinct types of odor receptors, which together allow you to detect thousands of different odors but do not provide specific location information.

- Each type of odor receptor in your nose requires its own unique gene in your DNA, which means you have hundreds of genes dedicated solely to creating olfactory receptors.
- Most of your sensory input goes first to the thalamus (the "switching station"), which then passes the information to the appropriate parts of the cerebrum. But in a break from this pattern, the olfactory nerves provide odor data to the amygdala before sending the information on to the thalamus. This means odor information gets a direct line to your emotion-processing center before your conscious brain even gets a chance to think about it.

MEMORY

Processes that occur entirely within the brain are especially intriguing to us. These are the processes we tend to focus on when we talk about the mind. It seems like a huge leap to talk about the anatomy, biochemistry, and electrical activity of the brain, and then to say that the mind is a consequence of it all. And it truly is a huge leap, in part because it glosses over the complexities that can emerge when trillions of small parts work in a connected manner. The mind within a human brain is a wonderful example of an emergent property. That said, we still don't have a full understanding of how it all works.

A key process that occurs entirely within the brain is memory, and for many people this feels like a much easier concept to grasp than the mind. Indeed, people often take memory for granted, thinking it's a simple and straightforward process, analogous to familiar technology. For example, we may say that memory is like a film or a video, with the memory written to the equivalent of a videotape or a hard drive in the brain. With reference to this model, we assume that a memory can be played back unchanged from when it was first stored in the brain. We further assume that a memory might slowly fade over time, like the fading of a photo, but that any detail that hasn't faded away is unchanged from the original memory.

But this isn't at all how memory actually works. The original experience on which the memory is based involved various bits of sensory input that triggered specific neurons in the brain. To recall a memory, we trigger some of those same neurons and then rely on the brain to *reconstruct* the memory, which is quite different from playing it back. This system has many weaknesses. First, our senses never capture *all* of the details from the original experience but instead focus on *specific* details. Some of those details are soon lost, failing to become part of an enduring memory. The neurons involved in each memory are also involved in many other memories, which means that each time we recall the memory, we tend to contaminate it with other associations even while reinforcing the strength of the altered memory with each reconstruction. Eventually the memory can diverge significantly from the original experience.

The upshot is that eyewitness testimony is often less reliable than we assume, even when the people testifying are absolutely certain they're telling the truth. This issue has particularly important consequences for our criminal justice system.

More Than the Sum of Its Parts

The more we look into the various components of the brain and the more we investigate how the brain works, the more we realize just how complex the system is. Thus, one of the biggest problems with the left-brained/right-brained concept is the reduction of the brain to a model with just two parts, along with the rather absurd idea that personality can be explained by such a simple model. After all, as you now know, each of the two hemispheres of the cerebrum is divided into four lobes, each with its own specialties. Within each lobe are zones that specialize in certain subtasks, and the brain divides any mental task into simultaneous subtasks that are allocated to various parts of the brain in both hemispheres.

Further, you now know that the brain has other crucial parts besides the cerebrum, including the cerebellum and the brain stem, as well as the thalamus, hypothalamus, hippocampus, pituitary gland, pineal gland, and amygdala. You understand how neurons work, and that a human brain has billions of neurons with trillions of connections. You realize not just the connectedness between all of these neurons but also how these neurons collect information from many different sense organs (both internal and external). You know that memories are not recordings but frail reconstructions of past sensory experiences that can easily be corrupted. And last but not least, by now you might appreciate that the enormous complexity of the brain—involving a huge number of parts, connections, and types of connections—can lead to the emergence of amazing properties, such as the conscious mind.

From a personal standpoint, I never could make much sense of the left-brained/right-brained model. Yes, I'm known for being logical, analytical, and methodical—and also for some distinctive quirks related to those attributes—but I'm also known for being an extremely creative problem solver. From my point of view, creativity is all about synthesis—finding ways of fitting disparate parts together in novel ways. But first I have to create a palette of disparate parts to work with. To do that, I employ analysis, which I see as the examination of complex systems to identify their component parts and how those parts work together to create a whole. Thus, I see creativity as equal parts analysis and synthesis, often in alternating waves. I use this approach in all creative endeavors, whether I'm creating a new algorithm to solve a difficult technical problem or a new musical composition that reflects many influences while also being unique.

Consequently, I've always rejected the idea that just because I'm highly analytical—and therefore "left-brained"—I cannot also be highly creative or "right-brained." We can be many things simultaneously; the complexity of the brain allows that to happen, even if our simplest popular models say otherwise.

Of course, your brain might have a different opinion on this matter—but don't let your left brain and your right brain get into a big fight over it!

12

Global Warming

F ew things intrigue me more than colorful presentations of data. As I flip through magazines or visit online articles about science, geography, or the world economy, I stop in my tracks the moment I spot a graph, chart, or diagram. I feel compelled to examine the graphic, because I'm incredibly curious as to what kind of data is being communicated and how it's being communicated. I then consider the implications of the infographic, and I consider what (if anything) is surprising about what I've just learned and whether the new data adds anything to my understanding of the world.

Although I'm attracted to all kinds of infographics, nothing grabs my attention as much as a map. I cannot resist studying any map I lay my eyes on. It can be a local map, a regional map, a country map, or a world map; I find all of them fascinating. It can be a highway map, a topographic map, a geologic map, a population density map, a map of ethnicities or languages, a map of vegetation zones, or a map related to economics and business; no matter, I find it irresistible. All of these maps impose a spatial representation of data on an abstract representation of the real world. If I visit a place depicted by a map I've studied, I can walk around in that real world using that spatially correlated data to better understand what I see with my own eyes. And even though I can't visit *all* of the places revealed to me in maps, the maps I study help me to understand people and places all over the world.

One category of maps has recently acquired a much greater urgency: maps that help us to analyze the climates of the world. We hear a lot these days about climate change, and we hear that the future consequences might be extremely serious. We also hear the term *global warming* mentioned frequently. But how do we know that the planet is warming and the climate is changing? How do we know that human activity is causing or exacerbating these changes? Why does an increase of carbon dioxide in the atmosphere have an effect on global temperatures? How quickly will these future changes occur and what will be the results? Will these changes actually be harmful, and if so, why? And if climate change is real, should we make a serious effort to stop it or should we just live with it?

What Does *Global Warming* Actually Mean?

The term *global warming* has gotten quite politicized in recent years, which is one of the reasons the term *climate change* is now preferred. The two terms are clearly related but have somewhat different meanings. Both terms suggest that the climates of the world are changing, which in turn suggests that climate maps could play a role in analyzing the situation. Now suppose all you wanted to do was to prove or disprove global warming. How could you do that? The answer depends in part on what the term *global warming* actually means.

To a scientist, the term suggests that the average temperature of the surface of the earth—including the atmosphere and oceans—is gradually increasing. But to some people, the word *global* carries a connotation of "everywhere, all the time." In fact, that's exactly how we use the word in certain other contexts, as a synonym for *universal*. And whether or not the word *global* causes any confusion, many people tend to conflate the concepts of weather and climate—that is, a snapshot in time versus a set of averages over time.

The result is that some people interpret the term *global warming* to mean a consistent and uniform increase in temperature at every point on the planet. In other words, if global warming is real, no point on Earth will ever experience a record cold temperature again. Following this logic, you could disprove the concept of global warming simply by pointing to a single record-cold temperature in a single location on a single day in a single winter. An even more convincing argument arises if the place where you live experiences a winter that's colder than the previous winter. This allows you to say that the entire concept of global warming is hogwash.

However, the term *global warming* actually refers to worldwide average temperatures, which are in fact rising. To put it another way, the real issue is the amount of heat retained in the global system (atmosphere, oceans, and land surfaces), not the temperature at a particular point at a particular time. This increase in heat energy produces a wide variety of effects—such as an increase in the severity of storms—that go beyond a mere rise in temperatures. These additional effects are another reason that *climate change* is now the preferred term.

In recent winters, whenever a major cold front (often associated with a "polar vortex") sweeps southward across the eastern United States, certain pundits say that this disproves the concept of global warming. But just because certain localities in the United States experience cold weather, it doesn't mean the entire world experiences the same thing at the same time. Consider a single day in a recent winter (December 29, 2017) on which a polar vortex caused very cold temperatures in parts of North America.

On that particular date, some places were experiencing temperatures much colder than average for that date, but other places were experiencing temperatures much warmer than average. Despite the chilly experience of people living in Washington DC and points north, the world as a whole was actually warmer than normal on that date. In fact, most of the Northern Hemisphere (including the western United States) was significantly warmer than normal for that date.

Notice that the concept of average temperature plays two distinct roles in this case. Each point on the globe can be said to have an average historical temperature for any particular date, and the deviation from that average can be recorded for that date. But to know whether the world as a whole is warmer or cooler than average, you need to average those deviations across the entire globe. On this day in December 2017, the world as a whole was 0.5 degree Celsius above the baseline average of 1979 to 2000. Furthermore, on this same day, the Northern Hemisphere was 0.9 degree C above the baseline. But to draw any real conclusions, you need to create a third average by averaging the daily global deviation across the entire year. This allows you to determine whether the world as a whole experienced a year that was hotter or colder than the historical average.

Note that using snowfall as a proxy for temperature simply doesn't work. Minneapolis is much colder than Buffalo, yet it gets only half as much snowfall. The South Pole, which has an incredibly frigid climate, receives far less snowfall than Minneapolis. So it certainly doesn't make sense to say, "My town received a record snowfall this year, so global warming isn't real." A large amount of snowfall in any location results from a great mass of warm, humid air crashing into a great mass of very cold air. It takes a lot of energy to evaporate water into the air and to move those big masses of air around the globe. Because of the greater-than-usual amount of energy in the atmosphere due to global warming, it makes perfect sense that some places will experience record snowfalls.

But Is the Earth Warming or Not?

The best way to determine whether the earth is warming is to compare the average worldwide temperature each year over a period of several decades. Our methods of data collection get better every year, which means the farther back in time we go, the poorer the data. For example, satellites now allow us to collect information about the surface temperature of oceans worldwide on a nearly continuous basis, but such data was not available 50 years ago. If we go back more than about 150 years, formal weather and climate data is rare and skimpy. Instead, we must rely on indirect information: historical mentions of

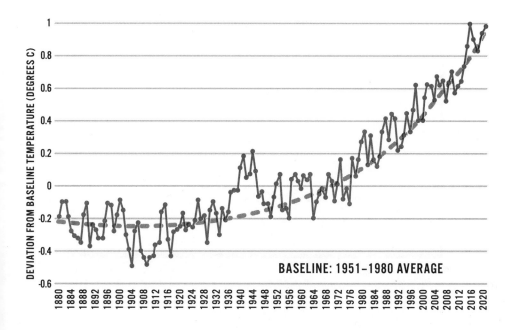

Average global land and ocean temperatures, 1880 to 2020

unusual weather, ice cores from glaciers, tree rings, and so on. Thus, it takes a lot of detective work to reveal climate patterns across a span of centuries. But even if we limit ourselves to a much shorter time frame, we can still see a clear pattern of global warming.

In the graph above, the dots represent the average global temperature each year from 1880 to 2020, compared to a baseline average (1951 to 1980). The dashed line is the smoothed trend line. The overall trend is clear, even though any specific year might be cooler than the year before. Notice that the trend was essentially flat for the years 1880 to 1920 and that the trend line has been getting progressively steeper ever since. This indicates two important things: (1) there really has been a trend of global warming in recent years, and (2) the rate of temperature increase appears to be increasing, meaning that the world is warming faster and faster.

We usually blame global warming on "greenhouse gases," especially carbon dioxide. So the next question is whether the levels of atmospheric CO_2 have actually increased in recent times. For this we have

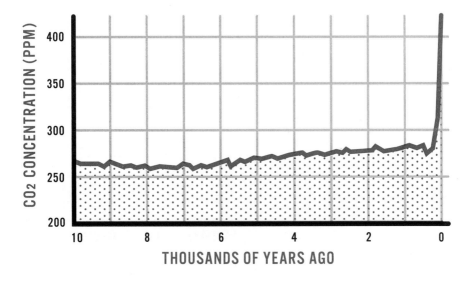

Atmospheric carbon dioxide over the past ten thousand years

excellent data. In addition to ongoing atmospheric measurements since 1958, we have thousands of years of data from analyzing air bubbles trapped in ancient layers of glaciers in Greenland and Antarctica.

You can see in the graph above that for most of the past ten thousand years, atmospheric CO_2 remained stable within a range of 260 to 285 ppm (parts per million). But in very recent times, the concentration has shot up to 415 ppm, and it continues to climb rapidly.

This leads to two crucial follow-up questions: (1) Although information from the past can appear to show a clear trend, it doesn't prove that the trend will continue. How can we reliably predict future global temperatures? (2) Although the increase in global temperatures correlates with the increase in atmospheric carbon dioxide, this isn't the only correlated factor, and besides, a correlation between two trends doesn't prove that one causes the other. So how can we reliably determine what's causing the increase in global temperatures?

These are big questions, and the complete answers are long. However, to answer either question, we first must answer this: How is it that greenhouse gases have the potential to raise the temperature of Earth? Or more simply: How do greenhouse gases trap heat? The answer to that question can help us answer the other two questions.

How Greenhouse Gases Trap Heat

I recently encountered a fascinating line of thought on the internet. Someone claimed that greenhouse gases are a myth, and he presented a simple argument to explain why. He said that if greenhouse gases actually trapped heat, they would cool the earth because they would prevent the heat of the sun from reaching it. To a certain extent, this is a great argument: it's simple, logical, and easy to understand. Unfortunately, the argument is flawed, and as a result the conclusions are false.

The flaws arise due to ambiguity in two key words: *heat* and *trapped*. The first problem is that this argument lumps several different forms of energy into a single term, *heat*. But the energy from the sun changes its form multiple times before it gets "trapped" by greenhouse gases. When speaking informally, we might refer to all these types of energy as heat, but this glosses over some crucial distinctions. The second problem is that this argument assumes that the word *trapped* means "blocked," but the word can also mean "captured," which has dramatically different implications.

To understand how greenhouse gases can warm the earth, you need to understand the following chain of logic:

1. The planet Earth constantly radiates energy into space, which is why our surroundings cool down at night.
2. For Earth to maintain a stable average temperature from year to year, the amount of energy coming from the sun and the amount returning to space must remain in balance.
3. The atmosphere has a huge effect on the energy coming in and going out, but different wavelengths of this energy are affected in different ways.

4. Changes in the earth's atmosphere, such as an increase in carbon dioxide, can easily upset the balance between incoming and outgoing energy, causing the overall temperature of the earth to change.

Let's examine a more detailed version of this explanation.

The first step in the process is for solar energy to travel from the sun to the earth. This is a long journey—about 93 million miles—but it takes only eight minutes because this energy travels at the speed of light. In fact, this energy *is* light. We sometimes say the sun gives off heat and light, because sunlight feels warm and it also allows us to see, but this is a description of human perceptions rather than a description of the energy itself. All of this energy is really just light, also called electromagnetic radiation. However, light can exist in a wide range of wavelengths, from gamma rays to radio waves, and only a limited subset of these wavelengths is visible to the human eye.

Sunlight isn't evenly distributed across the spectrum of light but is most intense in the visible part of the spectrum. Quite a bit of sunlight also occurs in the near infrared, which is the part of the infrared band closest to visible light. In fact, 95 percent of the energy that arrives from the sun is divided between the visible and infrared bands, with most of the other 5 percent in the near ultraviolet.

Upon reaching Earth's atmosphere, sunlight begins to encounter matter. This matter includes several kinds of gases, along with clouds and various types of particulate matter. These tiny particles, often called aerosols by climate scientists, are airborne bits of soot, dust, and other materials — some natural, and some caused by human activity. Each type of gas and each type of aerosol interacts differently with the incoming solar energy. Thus, sunlight has to run a gauntlet of obstructions before reaching the surface of the earth. These obstructions affect some wavelengths of sunlight more than others.

On average, about 30 percent of the incoming sunlight bounces back into space without being absorbed by anything. About 5 percent is backscattered by the atmosphere, 20 percent is reflected by clouds, and 5 percent is reflected by land and sea. The other 70 percent of the

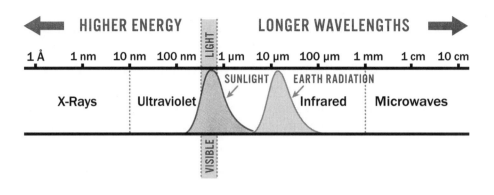

The spectrum of solar radiation reaching the earth's atmosphere and the spectrum of radiation given off by the earth

incoming sunlight is absorbed—about 50 percent by land and sea, and about 20 percent by atmosphere and clouds. These are averages for the entire earth across a full year, but weather conditions—especially with regard to cloud cover and snow cover—can strongly affect what happens at any single spot at any given time.

Any light reflected or scattered back into space continues to move at the speed of light and thus instantly leaves the vicinity of the earth. On the other hand, any light that is absorbed is no longer light. Instead it becomes heat (also called thermal energy), a completely different form of energy. Heat is (in essence) the motion of molecules of matter. The hotter the matter is, the more the molecules vibrate or move around. In contrast, the energy that travels from the sun to the earth does not involve molecules at all, just photons of light. When those photons reach the earth and are absorbed by matter, the photons disappear and the energy is converted into heat.

Now here is a really key point, one that most people don't think about. All objects give off radiant energy (light) all the time, even though we usually can't see it (except with special equipment). The amount of light given off and the wavelengths of that light both depend primarily on the temperature of the object. Hotter objects radiate a lot more energy than cooler objects, and the light from hot objects occurs at shorter wavelengths (which carry more energy). Any object cooler than about 900 degrees Fahrenheit (about 480 degrees Celsius) does not give off any visible light on the basis of its temperature. (When a cooler object gives off visible light, it's for a different reason altogether.) The objects that surround us on Earth primarily give off their radiant energy in the form of infrared (IR) radiation, which has longer wavelengths than visible light.

The upshot is that Earth constantly radiates IR energy back into space. During the day, the incoming solar radiation is much stronger than the outgoing radiation, for a net gain of energy on Earth. But at night the outgoing radiation never stops, so the surface of the earth cools down due to the loss of energy. In the preceding illustration, notice that solar radiation and Earth radiation have wavelengths that barely overlap.

Because of this difference in wavelengths, incoming and outgoing radiation react differently to the gases in the atmosphere. These gases are quite transparent to visible light, which is why the air looks clear and colorless. Clouds can obviously block some of the visible light, which is why we can't see through them and why it gets darker when a cloud passes overhead. Aerosols such as smoke and dust can dim the sky and make it hazy, again interfering with visible light. But the gases in the air don't block our vision, indicating they don't have much effect on the visible wavelengths of light.

In contrast, some atmospheric gases—such as water vapor and carbon dioxide—absorb a lot of energy in parts of the infrared band. This has a relatively small effect on incoming solar energy but a very large effect on Earth's outgoing radiation. If you look carefully at the lower half of the following illustration, you'll see that the atmosphere allows some wavelengths of light to pass through and absorbs other wavelengths. The top half is a reminder of what you saw in the previous illustration: the spectrum of solar radiation reaching the earth's atmosphere and the spectrum of radiation given off by the earth.

Notice that the short wavelengths on the left side of the illustration are completely absorbed by the atmosphere, including nearly all of the ultraviolet (except for a sliver of UV just beyond the visible spectrum). The long wavelengths on the right (the far IR) are also blocked. But a

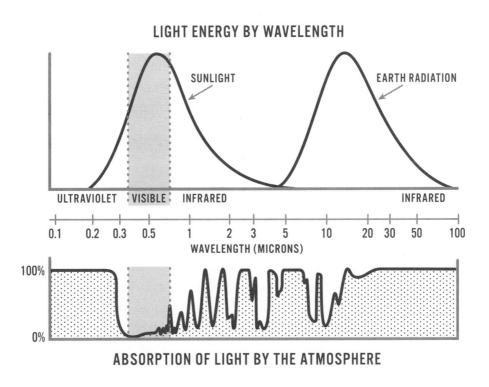

LIGHT ENERGY BY WAVELENGTH

SUNLIGHT

EARTH RADIATION

ULTRAVIOLET VISIBLE INFRARED   INFRARED

0.1   0.2   0.3   0.5   1   2   3   5   10   20 30   50   100

WAVELENGTH (MICRONS)

100%

0%

ABSORPTION OF LIGHT BY THE ATMOSPHERE

Absorption of light by the earth's atmosphere affects incoming and outgoing radiation in different ways

big window of transparency allows visible light to flood through from the sun to the earth without any significant impediment. Smaller windows allow certain wavelengths of IR to pass, but a lot of IR radiation is blocked, especially in the wavelengths that are given off by the earth.

The net result is that on average, 90 percent of the energy radiated by the surface of the earth is trapped before it can escape, absorbed by the atmosphere and clouds. The atmosphere is composed primarily of nitrogen and oxygen, but these are not the gases that block the IR radiation from escaping. Water vapor and carbon dioxide are the two gases that trap the most outgoing energy, while several other gases (such as methane) also play a role.

Now let's clarify the ambiguous term *trapped*. This word can result in several different mental models, some of which can lead to misconceptions. One possible model is to think of clouds and greenhouse gases as impermeable barriers, preventing energy from passing through and thus trapping it. This is the mental model used in the argument that if greenhouse gases really existed, they would prevent the "heat" of the sun from reaching the earth. It's true that clouds are a significant barrier, as we saw earlier, reflecting 20 percent of incoming solar energy back into space. This is because white surfaces reflect a lot of light. Landscapes covered in snow also reflect a lot of energy back into space.

Greenhouse gases, by contrast, *absorb* energy. They sop up—or trap—some of the light energy that passes through the atmosphere (mostly in the IR band). Unlike clouds, they don't reflect solar energy back into space. As greenhouse gases absorb energy, they heat up, keeping the energy right here in the atmosphere of the earth. These gases absorb only a small amount of the incoming solar energy but they have a huge effect on Earth's outgoing radiation, trapping it in Earth's atmosphere instead of allowing it to escape into space.

Balancing the Energy Budget

Here's another subtle but crucial point: for the temperature of the earth to stay essentially the same from year to year, the amount of incoming radiation must be exactly balanced by the amount of outgoing radiation. If the amounts are out of balance, the earth either warms up or cools off. This concept is sometimes called the energy budget of the earth.

Just as we itemized where the incoming solar energy goes (reflected by clouds, absorbed by the surface of the earth, and so on), we can analyze where the outgoing energy comes from:

- 30 percent consists of reflected or backscattered solar radiation that bounces away from Earth without ever being absorbed
- 10 percent is emitted by the surface of the earth
- 60 percent is emitted by the atmosphere and clouds

So only 10 percent of the radiation that returns to space is directly emitted from the earth's surface. The surface of the earth actually emits a huge amount of infrared energy, but most of this energy is absorbed by the atmosphere before it can escape. Likewise, the atmosphere emits a huge amount of IR energy, but half of it is sent downward toward the surface of the earth while part of the rest escapes into space. In other words, the earth's surface and its atmosphere are constantly trading energy, in large part by radiating infrared light back and forth. (Energy transfer also occurs via conduction, evaporation, and the movement of air.) To put it another way, the greenhouse gases in the atmosphere act like a blanket wrapped around the earth, keeping the planet warm.

Earlier I stated that water vapor is the greenhouse gas that most affects the outgoing infrared radiation. Thus, water vapor is an important factor in models of the earth's energy budget. Yet human activity has little direct effect on the total amount of water vapor in the atmosphere. We live on a water planet, two-thirds of which is covered by huge oceans of water, so the atmosphere has plenty of access to it.

On the other hand, human activity is having a dramatic effect on the amount of carbon dioxide in the atmosphere. CO_2, which is the second most important greenhouse gas, has a major absorption band right in the middle of the infrared wavelengths given off by the earth's surface and atmosphere. This means that carbon dioxide has a huge potential to decrease the amount of radiation that returns to space, thereby causing the planet to warm. This potential has been known for a long time, but predicting the exact result of any increase in carbon dioxide is devilishly difficult. We know enough to say that the current increase in CO_2 is indeed having a significant effect and that further increases will have further effects. Precise, reliable details are harder to come by, but each year we get better at measuring the current effects and predicting the future effects.

It's important to note that as the earth warms due to increasing CO_2 in the atmosphere, feedback loops affect other important factors. For example, warmer air holds more water vapor, which means that as the earth warms, increased water vapor in the atmosphere traps more of the outgoing radiation, amplifying the effect of the increased CO_2.

Another example: as the polar ice caps and glaciers shrink, the area covered by highly reflective snow and ice is reduced. Areas of newly exposed land and water absorb a lot more solar energy than they did before. Feedback loops such as these must be taken into consideration when building models that predict the warming effects of increased CO_2 in the atmosphere.

So now let's return to those two tough questions I raised earlier:

1. *How can we reliably say what's causing the increase in global temperatures?* The correlation between the rise of CO_2 in the atmosphere and the rise in global temperatures is helpful evidence, but it falls far short of proving a cause-and-effect relationship. Based on laboratory experiments, we have an excellent understanding of the effects of different types of greenhouse gases at various concentrations on different wavelengths of light. This provides strong evidence that increased amounts of greenhouse gases (such as carbon dioxide and methane) trap additional heat, causing an increase in global temperatures. Atmospheric measurements also provide strong evidence, showing that the atmosphere now traps more of the outgoing infrared radiation than it did a few decades ago. Other plausible explanations for the higher global temperatures have been investigated, and the most plausible explanation has proven to be the increase in greenhouse gases, providing an excellent fit with the data.

2. *How can we reliably predict what global temperatures will be in the future?* The simplest method to predict future temperatures is to extrapolate recent trend lines, but this is a crude approach. What is needed is a method that not only is more reliable but also allows us to consider various predictions of future concentrations of greenhouse gases, especially CO_2. Scientists meet this need primarily by creating ever-improving climate simulation models. One big issue in designing these models is how to apply the various relevant factors across the entire atmosphere while also accounting for the interactions among the atmosphere, the

oceans, and the land. Another issue is to ensure that the simulated results of these factors accurately reflect the actual data at hand. The present climate simulation models aren't perfect, but they're quite good, providing a great deal of insight as to what might happen in the future. Just as data collecting continues to improve over time, so do these models.

The Attempt to Halt Global Warming

Life on Earth depends on having water vapor in the air, which permits rain to fall. Life on Earth also depends on having carbon dioxide in the air, which allows the growth of plants, which feed all other forms of life (directly or indirectly). Furthermore, if not for greenhouse gases in the atmosphere, the world would be far too cold (especially at night). The problem is not that greenhouse gases are present in the air but that the quantities of certain gases (such as CO_2 and methane) are sharply increasing, which upsets the earth's balance between incoming and outgoing energy. When this balance is upset, it takes many years to reach a new equilibrium. If we somehow stabilized the amount of CO_2 in the atmosphere at today's level, the earth will still continue to warm for many years—perhaps even a century or more—until the equilibrium is restored.

But of course, humans continue to pour CO_2 into the air. When we finally get serious about becoming carbon neutral, it will likely take fifty years or more for the world to achieve that goal. And when humanity finally does become carbon neutral, this will not immediately lower the amount of CO_2 in the atmosphere; it will simply stop the level from continuing to rise. To reach a carbon-neutral state implies that we will have achieved several highly ambitious goals:

- generating all of the world's electricity without using coal or natural gas (although natural gas is a better alternative than coal)

- powering all of our transport vehicles—including cars, trucks, airplanes, and cargo ships—without using fossil fuels
- heating all of our homes and offices without using natural gas or fuel oil
- halting our destruction of rainforests and other carbon-storing ecosystems
- offsetting, modifying, or reducing our various nonfuel sources of carbon emissions, such as the manufacturing of cement and steel (which use carbon compounds as ingredients, partially converting them into CO_2 during the process)

It's going to be difficult to achieve carbon neutrality, but if we race toward a goal of reducing our CO_2 emissions by 90 percent, we can take a bit more time to deal with the last 10 percent.

Future global warming is likely to be far greater than what we have recently experienced. In the past hundred years, the world has warmed by only 1 degree Celsius (not quite 2 degrees Fahrenheit), and yet we can already see the effects in terms of warmer temperatures, shifting weather patterns, more severe storms and wildfires, melting glaciers, and rising sea levels. If we take immediate, decisive action to sharply reduce worldwide carbon emissions, we might be able to limit the future temperature increase to another 1 degree C. Under less rosy scenarios—which unfortunately are more likely—we can expect to see average temperatures increase by another 2 to 3 degrees C, and possibly even 4 degrees C.

Are We Facing Extinction— Or Is Everything Fine?

In the ongoing tug-of-war between people on either side of the global warming debate—those who want us to act decisively and those who don't—there are folks on both sides who promote highly exaggerated visions of the future. At one extreme are those who say if we don't act now, human civilization will come to an end and all life on Earth may go extinct. At the other extreme are those who say that a warming Earth will be a better and more pleasant place than today's world, so there's no need to do anything about it. Neither of these scenarios is a realistic assessment of what lies ahead.

Proponents of the need for decisive action paint a vision of a catastrophic future if we fail to act. In a sense they are probably right, but sometimes the argument is a bit misdirected. The real issue is not that a warmer world will be uninhabitable (which is almost certainly incorrect). Instead, the issue is that we face highly disruptive changes during the transition to a warmer world. Humans have come to rely on infrastructure, systems, and practices that require a stable and predictable world in order to function smoothly. Climate change is likely to upend all that, causing all sorts of problems that have the potential for disastrous outcomes.

Before I list the specific dangers we may face, let's first draw some lessons from the history of the earth. There have been times when the world was hotter than it is today and times when it was colder. Likewise, there have been times when sea levels were higher than today and times when sea levels were lower. For example, during the age of dinosaurs, the world was much warmer than today for long periods, and yet the world was full of life. So it is a mistake to suggest that a warmer world is antithetical to life.

Nearly all of the world's great extinctions have occurred during periods of dramatic climate change, indicating that these periods of change, rather than periods of warmth, pose the real danger. And while climate change can result in many species going extinct, it's also true that during the past 800,000 years, humans and countless other species successfully survived the dramatic changes in the latter part of the Ice Age. This span of time included eight glacial periods in which sheets of ice expanded across large areas of the earth, and eight interglacial periods in which the glaciers retreated, resulting in climates similar to those of recent times (the past 10,000 years). This means that in each cycle of about 100,000 years, a warm world changed to a much colder world and then back again. Each cycle included two global temperature swings of 4 to 7 degrees Celsius (with greater swings in certain places near the poles). The plants and animals of the world—including humans—had to adapt to these changes to survive, and entire ecosystems repeatedly had to migrate north to south and south to north. A lot of extinctions may have occurred during these changes, but a great deal of survival also went on.

Of course, another factor to consider is the rate of climate change, not just the amount. The current rate of human-induced climate change is extraordinarily fast, much faster than most of the climate changes in the past. Thus, the principal issue facing us is how to survive in such a rapidly changing world. Even when the time frame is not compressed, the transition from one climate regime to another is a period of high risk, likely to cause great difficulties for many species. The rapidity of the current changes will greatly enhance the risks. We need to be prepared for the difficulties that lie ahead.

The Future Effects of Climate Change

Predicting the future effects of global warming is an inexact science at best, mixing a limited amount of real science with a great deal of speculation about what might plausibly happen. A few things are certain—for example, that average global temperatures and sea levels will rise. We know that many aspects of the world will be heavily affected, such as weather patterns, crop production, population distribution, and of course, the natural ecosystems of the world. It's difficult to be much more precise than that while still guaranteeing accuracy. Our inability to predict the exact consequences of global warming underscores the risks we face: we're playing with fire, but we can't say with certainty how badly we'll get burned.

Still, we should look hard at the risks modern human society has created. During the glacial and interglacial periods of the Ice Age, the humans of the world (including *Homo sapiens* and other related species such as Neanderthals) consisted of small bands of hunter-gatherers. No infrastructure tied them to a particular locality, so they could simply move elsewhere if conditions changed and food became scarce. The state of humanity today is completely different. Now billions of people are spread across the globe, and we depend on our infrastructure to support this level of population. In particular, we rely on our agriculture and food distribution systems. We also live in permanent communities consisting of cities and towns, firmly rooted in specific locations.

In light of the current state of human society, let's take a look at the principal risks we're likely to face as a result of the next two centuries of climate change.

CHANGING WEATHER PATTERNS

Average temperatures will rise in most parts of the world, but this rise will be sharper in some places than in others. The increased energy in the atmosphere will increase the frequency and severity of major storms, including hurricanes, tornados, and winter storms. Changes in temperatures and atmospheric conditions are likely to cause changes in wind patterns and ocean currents. These changing patterns will, in many cases, have dramatic effects on where and when rain falls, which will have huge consequences. Think of the dramatic situation in many parts of the world when an El Niño year occurs and extrapolate this to an even wider range of effects in more places and with increased frequency. In some areas of the world, drought and higher temperatures will increase the risk of wildfire, a result we're already seeing.

SEA LEVEL RISE

Sea levels will rise during the next two hundred years, due to the melting of glacial ice and the thermal expansion of seawater. Predicting the amount and timing of this rise is complicated because periods of gradual rise might be punctuated by more sudden rises following partial collapses of the deep glacial ice over Antarctica and Greenland.

Hundreds of millions of people currently live in coastal cities and regions that are likely to be affected by sea level rise, and many will be displaced. Large numbers of homes and businesses may need to be abandoned. Coastal infrastructure such as ports, highways, and sewage systems will be severely affected.

FOOD SHORTAGES

Changing temperatures, changing rainfall patterns, and diminishing water in certain rivers will have a major impact on agricultural patterns around the world, affecting which crops can be grown in which places. This is likely to have a devastating impact on many subsistence farmers in poor countries. In richer countries with intensive farming practices, it will be necessary in many cases to adopt new crops or to move existing crops to new places in order to maintain current worldwide levels of food production. Certain agricultural pests might expand their ranges, presenting an additional challenge. If all of this is not handled effectively, food shortages could produce mass starvation in parts of the world.

WATER SHORTAGES

Rivers that depend heavily on annual snowmelt will see a sharp decline in summertime water flow. Rivers that drain areas of declining rainfall will also see falling water levels. Higher temperatures will increase evaporation rates, leading to additional water stress. But on rivers that drain areas of increasing rainfall, the excess water may cause damage to human water distribution infrastructure (such as dams and water treatment plants). Saltwater intrusion into some coastal aquifers, such as in south Florida, will disrupt local groundwater supplies. This range of effects will endanger water availability in many cities and agricultural regions.

POPULATION MIGRATION

All of these changes are likely to cause massive amounts of human migration over the next two centuries as people flee areas where their former modes of living have been made untenable. Some of the biggest waves of migration will originate from coastal areas and from areas where traditional forms of agriculture are failing. Much of this migration will cross international borders. Unless the world cooperates in the resettlement of these climate refugees, the toll of death and suffering will be staggering. But if we accept population movement as an obvious and reasonable response to climate change, we have the opportunity to minimize the amount of hardship.

ECOSYSTEM DAMAGE

Climate change will have a huge impact on natural ecosystems around the world, putting tremendous stress on vulnerable species that are already suffering from dramatic loss of habitat and declines in their population numbers. During the Ice Ages, when huge temperature swings occurred more slowly, entire ecosystems migrated with each swing, often moving hundreds of miles over a period of fifty thousand years. For example, at the peak of the glaciation, boreal forests now located in central Canada and high in the Rocky Mountains were instead located across the center of the United States. Coastal ecosystems abutting the ocean were located in places that are now completely covered by water and had to migrate to stay ahead of the inundation. The fragmented nature of today's remaining natural ecosystems will greatly hamper such migration in the future, and the rapid pace of change will make it even harder. To reduce the number of species that go extinct, we will need to take an active role in providing new locations for existing ecosystems, and in helping plant and animal species to make the move.

As for the oceans, increased atmospheric CO_2 will cause an increase in ocean acidity, which is detrimental to creatures that build structures from carbonate (as in seashells and coral reefs). Warmer water temperatures will have a major impact on which sea life can survive in which places. Furthermore, warmer water holds less oxygen than cooler water, and therefore a shortage of oxygen could seriously affect some sea life.

Adapting to the Changes

Although climate change is now occurring much more rapidly than during most previous episodes, the results will still take some time to unfold. The full effects of climate change will not be felt for more than a century, even though lesser amounts of change will be apparent much sooner than that. We have time to anticipate the changes and to prepare for them. Humans are quite adaptable, much more adaptable than most other species in the world. But to successfully adapt to this future world, and to preserve as many of our fellow species as possible, it would be helpful to predict these future changes in sufficient detail to map them. This brings us back around to the topic of maps.

Imagine a global, concerted effort to create detailed maps that depict the world as it will be a hundred years from now, even though global warming will probably continue beyond that date. We need maps that depict our best predictions of future climate zones, agricultural regions, and ecosystems. Such maps, frequently updated, would help us prepare for the future. We've already produced maps that offer predictions of future coastal flooding, based on various potential amounts of sea level rise; these other categories of maps would be equally valuable.

Personally, if I could put my hands on such maps, I would be glued to them for hours on end. I would view these maps over and over again, poring over them for details I hadn't yet noticed. I certainly don't expect to be around in a hundred or two hundred years, so I won't experience most of these changes myself. But the grandchildren and great-grandchildren of people I know will be here, and I'm intensely curious as to what their world will be like.

13

Epidemics and Pandemics

As a child and teenager, I had the habit of frequently sitting down by a set of encyclopedias to look something up and then remaining there for hours, reading one article after another. It was all driven by curiosity. A question would pop into my mind, so I'd walk over to the encyclopedia to find the answer. But the page that answered my question would pique my curiosity about related topics, so I'd open additional volumes. Soon I would be sitting on the floor surrounded by encyclopedia volumes, each of them open to one of those topics.

My dad heartily approved of my interest in the encyclopedia, so he offered me a hundred dollars if I would read the entire set from beginning to end. The idea had a certain abstract appeal to me—after all, it resembled the systematic way I ate the food on my plate. Furthermore, I was definitely tempted by the money. But once I started this project, I found it quite boring because it wasn't driven by curiosity. My purpose and joy in reading the encyclopedia were to answer the questions that piqued me on any given day, helping me to understand both the big picture and the related details. I was driven to pursue the linkages I perceived, uniting the various bits of information. So I found it extremely frustrating to read an article and then deny myself the ability to follow the related threads, due to a silly mandate to strictly follow the alphabetical organization of the encyclopedia. I soon abandoned the project, although I kept reading the encyclopedia every time I had a question.

Many years have passed since my childhood, but my reading is still largely driven by the questions I have. Like many people, in 2020 I had a lot of questions about epidemics and pandemics. Some of these questions were about the definitions of words, but other questions went much deeper, inquiring into the meaning and implications of the disjointed bits of information that came flooding to me in the media. I kept looking for the linkages that would allow me to assemble a big picture while also filling in some of the details.

My questions began at the broadest level: What does the word *pandemic* actually mean? The term is obviously related to the similar word *epidemic*. The Greek root *pan* means "all" or "everywhere," so I assumed that a pandemic must be an epidemic that has gone global. But then I kept running into the phrase *global pandemic*, which seems like a term invented by the Department of Redundancy Department. I eventually learned that an epidemic doesn't have to be global to be a pandemic, but it does have to be more widespread than a typical epidemic. On the other hand, anything called an epidemic has already spread significantly; otherwise, it would merely be called an outbreak. So where do we draw the lines among these terms?

I've run across other issues that make it tricky to figure out what the word *epidemic* really means. I keep seeing headlines that shout phrases such as "an epidemic of drug overdoses," "an epidemic of suicides," "an epidemic of obesity," and "an epidemic of fear." Do all of these phrases simply represent metaphors, or are any of these afflictions actual epidemics? If these are just metaphors, does the metaphor merely imply that something unpleasant is increasing in frequency? (I seldom see references to "an epidemic of happiness" or "an epidemic of generosity.") Does the word *epidemic* imply a linkage between the individual cases, suggesting that they spring from a single cause? Does it imply that the malady is contagious?

Of course, we most often associate the words *epidemic* and *pandemic* with viral and bacterial diseases that spread rapidly. We also associate the words with a lot of deaths. Does a disease have to be deadly for an outbreak to be called an epidemic or a pandemic? Why do these diseases suddenly appear and spread so rapidly? How can we protect ourselves from these diseases? Can we predict pandemics before they happen? And finally, just out of curiosity, what have been the most notable pandemics of the past few centuries? Are these ancient diseases gone for good, or can they return?

Yes, I'm flooded with questions, and I imagine you must be too. So let's see if we can come up with some good answers.

Defining the Terms

Let's start by figuring out the actual meanings of the terms *pandemic*, *epidemic*, and *outbreak*, and the differences among them. If you look up *epidemic* in several different sources, you'll get slightly different definitions, but the core idea is that an epidemic is a disease that affects a large number of people at roughly the same time. Some (but not all) of the definitions insist that it must be an *infectious* disease— that is, a disease spread by an organism like a virus or a bacterium. (This narrower meaning would rule out such maladies as alcoholism, obesity, and most cancers.) Many of the definitions include a geographical aspect—in other words, *epidemic* implies a high incidence of a specific disease in a specific community, region, or country. And finally, another frequent theme is that the word implies a sharp increase in the number of cases. If a high incidence of the disease holds steady for many years, you would say that the disease is *endemic* to the place, rather than calling it an epidemic.

Okay, I think that gives us a handle on the word *epidemic*. Let's move on to *pandemic* and *outbreak*. The word *pandemic* is used when an epidemic has spread to many different countries, usually on multiple continents. Once a disease has become so widespread, it's much harder to control. An *outbreak* is the early stage of what might eventually become an epidemic, when the number of people affected is still small and so is the geographic area. Some outbreaks die out by themselves, but others go on to become epidemics, which can sometimes become pandemics. Thus, it's important to identify outbreaks early, to control them before they spread.

Scientists who study outbreaks, epidemics, and pandemics are called epidemiologists, and their field of study is called epidemiology. Whenever such an event occurs, epidemiologists rush to understand as much as possible about the disease—as quickly as possible—in order to make recommendations about how to control the spread of the disease. Public health officials can then act on these recommendations. Key questions to answer include: Is the cause of this disease an infectious organism or something else? If so, which organism causes it? How does the disease spread? How contagious is the disease? How lethal is the disease? Are some people affected by the disease more than others? Other medical researchers, outside of the field of epidemiology, then pursue closely related questions, such as: How can we develop a treatment for this disease? How safe and effective are the candidate treatments? How can we develop a vaccine for this disease? How safe and effective are the candidate vaccines?

To answer all these questions, the response to an epidemic typically involves researchers with a wide range of backgrounds and specialties. Medical doctors and field biologists visit the sites of outbreaks to collect samples and data onsite (in addition to treating the afflicted and working to stop the spread of the disease). These samples and data provide crucial information regarding the initial cases, along with clues about the origin of the disease. Laboratory researchers study the field samples to identify any disease organisms found in the samples. Scientists study the data collected from field studies and lab studies, aggregating and correlating the data to make sense of it and to draw conclusions from it. Computer modelers work with the accumulated data in order to predict the future spread of the disease, based on various scenarios regarding what steps we might take to control the disease.

But that's not all. Still other scientists work to decipher the DNA of the disease organism, which is crucial to enable other steps, including the creation of a vaccine, understanding where the disease organism came from, and figuring out how quickly the organism is mutating into multiple strains. As the disease spreads to become an epidemic, health officials in the affected localities collect public health data that is essential to understanding the progress and the threat of the disease. Still other people measure the effects of various countermeasures that are employed. All of the new data helps to improve the ongoing computer modeling, as well as the recommendations for controlling the disease.

The Ambiguous Meaning of *Deadly*

Note that the definition of *epidemic* does not state that the disease must be deadly. For example, a particular community might experience an epidemic of skin rashes caused by a fungus, or an epidemic of head lice. However, the deadly epidemics are the ones that get the greatest attention, for good reason. Whenever we hear of a dangerous epidemic, we all want to know if we're personally in danger. We want to assess our degree of risk along with the nature of that risk, and also the risk to people we know.

In the popular media, the common way to express the deadliness of a disease is to count the number of deaths. Likewise, when we look at epidemics of the past, we compare them by estimating how many people died. But when we look at our future risk from a specific disease, we need a different approach. I usually think of the risk as having two separate parts: (1) How likely am I to contract the disease? (2) How likely am I to die if I do get it? The first factor corresponds to contagiousness, while the second factor corresponds to lethality. You need to consider both factors to assess the danger of a particular disease. (Other factors, such as the likelihood of permanent damage to your health, could also affect your assessment of the risk.)

Note that I'm using informal terms here. What I'm calling lethality is what a scientist would call the infection fatality rate (IFR), computed by dividing the number of deaths by the number of infections. (The related concept of case fatality rate, or CFR, is computed using the number of proven cases, rather than an estimate of all infected people.) What I'm calling contagiousness can also be called transmissibility. The actual rate of transmission (which we try to lower by following appropriate health precautions) is often expressed by epidemiologists as the basic reproduction number ($R0$, pronounced "R-naught").

The surprising fact is that a disease with a low rate of lethality can often kill a lot more people than a highly lethal disease. This point can be illustrated by comparing Ebola to the seasonal flu. Ebola is a disease caused by any of several different ebolaviruses. Since the disease was first identified in 1976, it has killed not quite fifteen thousand people worldwide. By comparison, influenza (flu), which is caused by any of several different strains of flu virus, kills an average of thirty-five thousand people each year in the United States alone, and about ten times that many worldwide. And yet we find Ebola to be much more frightening than flu. Why is that?

The main reason for our indifference to flu is the low fatality rate, compared to a disease like Ebola. About half of the people who contract Ebola die from it, meaning it has a fatality rate of about 50 percent (depending on the species of the virus and the locally available treatments). In contrast, seasonal flu typically kills fewer than one out of a thousand infected people—a lethality rate of 0.1 percent or less (again depending on several factors). So which disease should we declare to be more deadly: the one that kills more people or the one that kills a higher percentage of the people it infects? The answer clearly depends on how you define *deadly*, but I'd argue that the total number of deaths is the better measure, and that's the definition I'll use here.

Two other factors help to explain our indifference to the risk of flu. One is that flu is quite familiar; it comes back every winter, affecting a roughly similar number of people each year. Thus, in most years the disease is endemic rather than an epidemic, and we're more willing to ignore risks that arise from familiar causes. (But when new varieties of flu appear, they can sometimes become deadly pandemics.) Another factor is that flu deaths mainly occur in elderly people, very young children, and people in poor health. If you are a healthy young adult, you might perceive your risk of dying from flu as exceedingly small, in sharp contrast to the risk of dying from Ebola should you ever catch it.

This ambiguity in the definition of *deadly* has played a role in our varied perceptions of COVID-19, the novel coronavirus that suddenly caused a worldwide pandemic beginning in early 2020. In the first twelve months after the disease began its rapid spread, it had already killed about two million people across the globe. Judging by the body count, COVID-19 is far more deadly than the seasonal flu. But is it actually more lethal or simply more contagious? A commonly quoted estimate put the infection fatality rate from COVID-19 at around 0.6 percent—extremely low compared to Ebola but much higher than a typical seasonal flu. However, COVID-19 is easily transmitted through the air from person to person. And when the disease first spread around the world, no one had previously acquired immunity to it, in contrast to the seasonal flu. Thus, a mix of several factors caused COVID-19 to be a far more deadly disease. The core issue is that so many people caught the disease so quickly. Under such circumstances, even a very low death rate can produce a great number of deaths.

Why Epidemics Suddenly Appear and Spread

The two most significant pandemics of the past few decades have been COVID-19 and AIDS (caused by HIV). Other outbreaks that have received a lot of media coverage include Ebola, SARS, Zika, the H1N1 (swine) flu, MERS, and Marburg. These particular diseases have two things in common: (1) all are caused by viruses, and (2) all crossed over to humans from other animals. Not all diseases are caused by viruses; plenty of serious diseases (such as malaria and tuberculosis) are caused by other kinds of organisms. However, whenever we hear in the news of a big outbreak of a recently recognized disease, it's usually caused by a virus that normally resides in other animals but has made the jump to humans.

Birds, bats, and rodents are the animals that harbor most of the viruses that cross over to people. For example, birds are the principal source of flu viruses, and bats are the principal source of coronaviruses. The tricky part is that in many cases, such diseases are first passed to some other animal before being passed to us. For example, swine flu originated in pigs as a hybrid of three different flu viruses, but birds provided the reservoir of original viruses. SARS also took a two-step route to reach us (and COVID-19 probably did too).

The amazing thing about all of this is that the original host species is often barely affected by the disease. The birds, bats, and rodents carrying the diseases seldom die from the common forms of these infections. Instead, the virus has evolved to the point that the host can stay relatively healthy, which is advantageous for the virus too. A host that stays alive is more likely to provide a long-term home for the virus as well as providing additional chances to spread the disease. Evolution on the part of the host species can likewise result in an increased tolerance to an endemic virus, thereby contributing to this ability to get along.

Although these viruses are well adapted to their normal hosts, they are often poorly adapted for infecting humans. As a result, one of two issues often arises, limiting the ability of the virus to establish a permanent presence in a human population. First, the virus might be unable to spread beyond the person it infects. (This is the case with the deadly H5N1 avian flu, which can only be acquired from a bird, not from another human.) Second, the virus might be overly lethal in humans, making it less likely that the virus will spread before the victim dies. However, genetic variations frequently appear in viruses, and sometimes such a variant is better adapted to its new host—either more contagious, less deadly, or both. This new strain might then be able to spread quickly in humans.

Another factor contributing to the evolution of viruses is a dual infection, when a host cell is invaded by two related viruses simultaneously. This means that both viruses have injected their genetic material into the same cell, directing the cell to make additional copies of both. In this circumstance, the cell sometimes produces a hybrid virus, combining genes from each of the two viruses. In fact, this is how swine flu arose, when *three* different species of flu combined in the cells of pigs. As this new flu spread around the world, scientists were worried that it would combine again but this time with a far deadlier strain of flu. Luckily, this did not happen.

Predicting and Preventing Pandemics

Is it possible to predict pandemics before they occur? Predicting pandemics is somewhat similar to predicting earthquakes. We can't say exactly when the next major pandemic will occur or where it will start. But we know for sure that it will come, and that other pandemics will follow. In most cases, the outbreak will result from a virus crossing over to humans from some other species of animal. We know that such crossovers occur most often in eastern Asia and central Africa— in part because many of the wild animals that carry such viruses live there—but we also know that some epidemics will begin outside of these two zones. If our early warning systems are effective, we will detect outbreaks soon after they occur, allowing us to warn the world to take immediate action.

Preventing all viral crossovers is not possible, although we can reduce the frequency by training people to avoid risky practices (such as eating animals that have been found dead) and discouraging crowded wildlife markets (where viruses can easily pass from one species to another). But because we cannot prevent all viral outbreaks, we must also rely on the prompt and effective containment of individual outbreaks soon after they occur. Notice that this actually requires two distinct processes. First, a system needs to be in place for quickly realizing when a serious new disease has broken out, anywhere in the world but with special focus on the places where new diseases are most likely to emerge. Second, a rapid and effective response must be launched when a new outbreak is detected (or suspected).

At the first sign of a dangerous new outbreak, local authorities must establish a strict quarantine of the affected area. (In practice, it tends to be easier to quarantine a small, isolated rural area than a densely populated urban area.) Then, in association with national and international experts, authorities need to answer as many key questions as possible, in the briefest possible time: What is the disease organism? How is it spreading? How contagious is it, and how lethal is it? What is the genetic sequence of its DNA?

Unfortunately, simply placing an area under quarantine is seldom enough to contain an outbreak. Some of the infected people might have spread the disease before the quarantine took effect. Other people might sneak out of the quarantine area. To effectively halt the spread of the disease, contact tracing is usually necessary. This means that everyone who has gotten sick must be identified, and the recent travels and interactions of that person must be traced. Everyone who came in contact with the sick person must be tracked down and quarantined or tested for the disease. Of course, all of this effort must occur simultaneously with the medical efforts to safely treat all the people afflicted with the disease.

Countries, states, cities, and hospitals need to prepare in advance for future epidemics. The first level of preparation is to collect supplies and establish clear procedures for dealing with a future outbreak. A second level of preparation should occur immediately after warnings that a new epidemic has begun. This includes the manufacture of additional supplies that are likely to be needed for dealing with this specific disease. Authorities at various levels of government and public health care must quickly agree on procedures for patient transport and care, triaging cases, quarantining the exposed, testing for the disease, and tracing contacts, in addition to planning the logistics for the ongoing production, purchasing, and distribution of additional medical supplies.

This raises an important question: How do we decide whether a specific epidemic is worth an all-out effort to contain it? Few people would question a massive effort to contain an outbreak of a disease with a 50 percent fatality rate (such as Ebola). But such a consensus is far less likely when a disease has an infection fatality rate of less than 1 percent (as with COVID-19), no matter how contagious the disease is. Furthermore, maintaining a consensus is far easier when a deadly outbreak can be quickly contained. If the disease persists in a population for many months, people get weary of the measures to control it. The upshot can be a lose-lose situation: we lose if we let the disease get out of hand, and we lose if the countermeasures stifle the economy. Finding the optimal middle path is tricky.

With regard to COVID-19, different countries chose different approaches to deal with the epidemic. Countries that promptly implemented strict and effective measures against COVID-19 (such as New Zealand, Taiwan, and South Korea) quickly got the disease under control, keeping the number of sick and dead quite low, which allowed their economies to safely reopen much sooner than would otherwise have been possible. But after the initial success, it was necessary to continue with aggressive testing, contract tracing, quarantining, and other measures in order to prevent a resurgence of the disease.

Such success stories don't eliminate the dilemma of deciding which future epidemics ought to trigger maximum countermeasures. Suppose a new viral epidemic breaks out two years from now, but the virus is not quite as contagious as COVID-19 and is only half as lethal. What measures should we take to fight it? Should we go into full lockdown? Should we continue business as usual but get really serious about testing and contact tracing? Or should we do nothing other than warn the public? These are not easy questions to answer, but in making these decisions we should pay careful attention to the predictions of epidemiologists, based on computer models that reflect various combinations of countermeasures. And we should not forget that the manner in which we respond can have a huge effect on the ultimate outcome, including the total number of deaths.

How Can You Protect Yourself During an Epidemic?

Suppose you find yourself in the midst of another epidemic. What is the best way to protect yourself? Let's assume we're talking about an infectious disease, spread by a disease organism such as a virus, and that a vaccine is not yet available. In that case, the first question is how the infection is spread. Is the disease contagious—in other words, does the disease spread via human-to-human contact? If so, how? If not, is it spread by insects, such as mosquitos? Once we understand the principal mechanisms by which a particular disease spreads, we can protect ourselves by obstructing those mechanisms, thereby breaking the cycle of transmission.

For example, viral respiratory diseases such as COVID-19 tend to spread primarily by little droplets released into the air when an infected person coughs or sneezes (and to a lesser degree when an infected person talks, sings, or shouts). These airborne droplets are teeming with virus particles. Other people can contract the disease after inhaling those droplets. For some respiratory diseases, such as flu, a related source of transmission may be the infected droplets that fall onto surfaces. When people touch an infected surface and get the virus particles on their hands, then touch their eyes, nose, or mouth, they unwittingly introduce the virus into their bodies. The safety recommendations promoted by health officials—wearing a mask, avoiding crowds of people, washing hands frequently—are all designed to break the transmission cycle, thereby reducing the spread of the disease.

Not all contagious diseases are respiratory diseases; some are primarily spread by mechanisms other than droplets coughed or sneezed into the air. For example, Ebola is spread primarily by body fluids, such as blood, diarrheal feces, and vomit. Thus, the disease typically spreads to the people who take care of Ebola patients. If the patient dies, as so many do, the body remains highly infectious, putting at risk the people who prepare the body for burial. When an outbreak of Ebola occurs, public health officials focus on breaking these modes of transmission by promoting safe methods of treating patients and handling bodies.

Quite a few infectious diseases, such as malaria and Zika, are spread primarily through the bites of mosquitos. The public health measures to control these diseases tend to focus on reducing the local populations of mosquitos and reducing the likelihood of being bitten by the mosquitos that remain. Insect repellents and mosquito netting are especially useful tools for breaking the cycle. In the case of yellow fever, another disease spread by mosquitos, a vaccine is available. For malaria, several prophylactic drugs are available that greatly reduce the odds of getting infected by the disease should you get bitten by malaria-carrying mosquitos.

Unfortunately, cures don't exist for most epidemic diseases, making a focus on prevention the best option. In particular, viral diseases and diseases caused by protists (such as malaria) are usually quite difficult to cure. That said, some of these diseases can be *managed* (but not cured) with a careful mix of drugs. Modern drugs have dramatically lowered the death rate from AIDS and can often improve the health of malaria victims. Still, virtually all of these patients will continue to be afflicted with their disease for the rest of their lives. Other diseases, including Ebola and COVID-19, will eventually disappear from a victim's body. In these cases, certain drugs can increase the odds that the person will survive long enough for this to happen, and some drugs might shorten the course of the disease by a few days.

Notable Pandemics of the Past

What are some of most notable pandemics of the past? We know that several pandemics occurred in the empires of the ancient world, especially the Roman Empire. However, we don't know for certain which diseases caused any of these pandemics. In contrast, several of the later pandemics that swept through Europe and the Middle East were apparently the work of bubonic plague, identifiable by the distinctive symptoms.

Bubonic plague is a bacterial disease typically transmitted to humans through the bites of fleas. The lymph nodes nearest the bite swell and become painful, which is the source of the name of the disease (a *bubo* being an inflamed and enlarged lymph gland). The fleas themselves usually pick up the bacteria from biting infected rodents. Bubonic plague is a serious disease, and untreated cases have a fatality rate in excess of 50 percent. Of course, there was no effective treatment until the invention of antibiotics less than a century ago. But even when the disease is treated with modern medicines, the death rate is around 10 percent.

We know of several epidemics of bubonic plague in centuries past, three of which are considered to have been pandemics. The first such pandemic struck the Byzantine Empire in the years 541 and 542, causing many deaths. A much bigger pandemic—called the Black Death—swept through Europe in the years 1347 to 1351, killing an estimated fifty million people. (While it's generally accepted that the Black Death was caused by bubonic plague, some dispute this.) Smaller epidemics (but still quite deadly) popped up in later centuries, most notably from 1720 to 1723. Another pandemic in the mid-1800s originated in China and eventually spread to locations around the world, killing tens of millions of people. Even the United States was affected by this third pandemic, primarily on the West Coast. Bubonic plague still exists, but in current times it tends to infect just a few hundred people each year.

While bubonic plague was the most feared of the deadly diseases that rampaged through Europe and elsewhere, it was certainly not the only killer. Crowded conditions in dirty cities were especially amenable to the spread of diseases including smallpox, typhus, influenza, diphtheria, leprosy, malaria, measles, cholera, anthrax, scarlet fever, tuberculosis, and whooping cough. After centuries of exposure to these diseases, people in Europe acquired a limited resistance—in other words, they gradually became less likely to die when infected. But with the "discovery" of the New World, conquerors and settlers from Europe soon introduced all these diseases to Native American populations, who had no such resistance. Smallpox, in particular, was devastating to the native people of the Western Hemisphere, but several of the other introduced diseases were also major killers. Tens of millions of Native Americans soon died from these diseases—at least half of the Indigenous population and perhaps as much as 90 percent. These epidemics eventually affected the entire hemisphere, the entirety of the Americas, producing enormous consequences.

Smallpox has the rare characteristic of occurring only in humans. In other words, there is no reservoir of the virus lurking in other species of animals. Thus, it was conceivable that the virus—one of the deadliest diseases to afflict humanity—could eventually be wiped out. Following a massive international campaign of vaccinations and quarantines that went on for decades, the virus was declared to be completely eliminated from the human population in 1980. (However, the virus still exists, because some specimens were preserved in secure laboratories.)

With the decline of bubonic plague and the elimination of smallpox, influenza (flu) took over the role as the worst source of pandemics. This may seem counterintuitive, because in most years the seasonal flu has a lethality of only about 0.1 percent (or even less). Of course, flu is highly contagious, which means that even 0.1 percent can equal a great number of deaths. However, new variants of flu constantly arise, and every now and then a variant appears that's especially deadly. For example, the flu pandemic of 1889–90 killed an estimated one million people. But the "big one" was the infamous Spanish flu pandemic of 1918–20, which killed an estimated one hundred million people around the world. This pandemic was especially baffling because so many of the victims were healthy young adults—in other words, the type of people who would normally survive a case of flu. Luckily, that particular strain of flu soon died out.

In the past century, the world has not experienced another pandemic as deadly as the Spanish flu, but there have been subsequent flu pandemics. The worst, in terms of number of deaths, was the Asian flu pandemic of 1957–58, which killed more than a million people worldwide. The 1968 Hong Kong flu pandemic was a close second, killing about a million people. In a typical year, between 250,000 and 500,000 people die from flu worldwide, still quite a large number when you consider that those deaths occur from a normal seasonal flu.

Outbreaks and Epidemics in Recent Years

Let's now focus on recent outbreaks and epidemics (those of the past fifty years), which can be divided into several distinct categories:

- **INFLUENZA** Flu continues to be a serious health issue every year, killing many people, primarily the elderly and the very young. Each year epidemiologists try to predict the principal strains of flu that will sweep the world the following winter so that the annual vaccine will offer protection against the correct variants. A related worry is that a more deadly strain of flu will again appear and spread around the world, causing far more deaths than usual. In fact, any new strain of flu in humans poses an increased risk, because no one has yet acquired a natural immunity from previous exposure. We dodged a bullet with the H1N1 swine flu pandemic in 2009–10, because this highly contagious new variant turned out to be no more deadly than a typical seasonal flu. However, the day will eventually come when our luck runs out, and a much more lethal strain of flu becomes a pandemic. Although birds serve as the principal reservoir for flu viruses, pigs are capable of catching both avian and human variants, making them an ideal bridge for infecting humans with a new flu virus. Therefore, as part of a worldwide early warning system, epidemiologists keep a close eye on any flu epidemics that suddenly appear in birds or pigs.

- **CORONAVIRUSES** Until the COVID-19 pandemic, most people had never heard of coronavirus, but in fact several coronavirus diseases have infected humans in recent years. The two most lethal have been SARS (2002–04) and MERS (2012 until the present), with SARS killing 11 percent of infected people and MERS killing 35 percent. The SARS epidemic came close to becoming a pandemic but was contained after more than eight thousand cases emerged worldwide. The virus that causes COVID-19 is genetically quite similar to SARS but more contagious and considerably less lethal. (The official name for the COVID-19 virus is SARS-CoV-2). That said, COVID-19 has proven to be far more deadly than SARS because many millions of people have become infected. Four other species of coronavirus have also spread around the world, but none of these four are deadly; they merely produce symptoms like that of a common cold. In fact, perhaps 20 percent of all cases of common cold are caused by those four mild coronavirus species.

- **HIV/AIDS** AIDS (acquired immunodeficiency syndrome), the disease caused by HIV (human immunodeficiency virus), first appeared in Africa, crossing over from chimpanzees to humans. The disease remained under the radar until after it began to spread around the world in the 1970s, but the viral agent producing the disease was not identified until 1983. Without treatment, most AIDS victims will eventually die from the consequences of the disease, usually living for several years before that happens. Although no cure for AIDS exists, the disease can now be effectively treated with a powerful "cocktail" of drugs. Because these drugs don't completely destroy the virus, the patient must continue taking the drugs for life. Since the beginning of the pandemic nearly fifty years ago, about seventy-five million people have been infected with AIDS and about thirty-three million have died. Although the death toll has declined dramatically, more than a half million people still die from the disease every year.

- **EBOLA** Of the five known species of ebolavirus, four have caused fatal outbreaks in humans. The most dangerous of the four is Zaire ebolavirus, responsible for most of the outbreaks and by far the most deaths. Perhaps a dozen separate outbreaks of the disease have occurred, dating back to the 1970s, and each outbreak was likely caused by a separate case of the disease crossing over to humans. Most outbreaks of Ebola are contained with no more than a few hundred deaths. However, the disease is quite capable of causing a major epidemic, as shown by the outbreak in Guinea, Sierra Leone, and Liberia (2013–16) that killed more than eleven thousand people.

- **TUBERCULOSIS** Tuberculosis is an endemic disease in many parts of the world, killing about 1.5 million people a year, which in most years is more than any other infectious disease. (However, in 2020 COVID-19 leapt ahead of tuberculosis.) Unlike many of the diseases discussed in this chapter, tuberculosis is a bacterial infection rather than a viral infection.

- **DISEASES SPREAD BY MOSQUITOS** Malaria continues to be a major world health threat, killing a half million people each year and permanently damaging the health of many more. The list of diseases spread by mosquitos is surprisingly long, including many common diseases and many rare diseases. Other than malaria, the mosquito-borne diseases most often in the news are Zika virus, West Nile virus, Chikungunya virus, dengue, and yellow fever. Just a few years ago, a massive Zika outbreak grew into a pandemic, hitting Brazil especially hard. This disease became notorious not because of a high death rate but because of the birth defects it often causes when pregnant women become infected. The good news is that the pandemic is over, with no major outbreaks anywhere (as of early 2021). The bad news is that the disease could easily return.

- **DISEASES SPREAD BY RODENTS** We no longer see great epidemics of bubonic plague, but several other diseases spread by rodents are still noteworthy. One of the most deadly is Lassa fever, which kills about five thousand people a year in western Africa, although perhaps only 2 percent or 3 percent of cases are fatal. (In other words, the disease is fairly common.) A relatively rare disease that has gotten a lot of press is hantavirus, which humans can acquire by coming into contact with the excrement of wild native rodents. In the United States, more than seven hundred cases were reported over a twenty-three-year period, with 36 percent of the cases resulting in death.

So which of these infectious diseases have been the deadliest killers of the new millennium? In the first decade after 2000, AIDS continued to kill more people than any other infectious disease, until increasingly effective treatments brought the numbers down. In the second decade of the millennium, tuberculosis took the top spot as the deadliest infectious disease in the world. But then in 2020, a brand-new disease named COVID-19 suddenly became the leading killer.

At the international level, the leading organization for monitoring outbreaks and epidemics is the World Health Organization (WHO). Within the United States, the leading organization is the CDC (Centers for Disease Control and Prevention). It's worth noting that the letters *CDC* originally stood for Communicable Disease Center, which signaled an emphasis on infectious and contagious diseases. But now the organization has a broader mandate, focusing on any sort of risk to public health. Epidemiologists take an interest in any geographic cluster of diseases, so it might be more than just a metaphor to speak of an epidemic of cancer or an epidemic of liver failure. Even if no infectious organism is involved, a noninfectious agent may be playing a role, such as chemicals in the drinking water or the habit of smoking cigarettes. If so, epidemiologists want to know!

Early in this chapter I raised a great number of questions, and I think I've managed to answer most of them. Doing the research for this chapter reminded me of those days long ago, sitting on the floor surrounded by all those encyclopedia volumes. But these days I can do the work while remaining seated at my desk, with a couple dozen tabs open in my web browser plus a stack of books from my favorite science authors tottering next to my keyboard. Bouncing back and forth among all these sources, I was able to pursue enough threads to assemble a big picture, backed by a collection of supporting details. And I did all this without having to sit for hours on the floor—which is certainly a good thing, because I don't think I could do that anymore!

FURTHER READING

If you would like to read more about the topics covered in this book, along with other interesting science topics, I recommend any of the following books.

Brusatte, Steve. *The Rise and Fall of the Dinosaurs: A New History of a Lost World*. New York: William Morrow, 2018.

Bryson, Bill. *A Short History of Nearly Everything*. New York: Broadway Books, 2003.

Dunn, Rob. *Never Home Alone*. New York: Basic Books, 2018.

Henshaw, John M. *A Tour of the Senses: How Your Brain Interprets the World*. Baltimore, MD: Johns Hopkins University Press, 2012.

Henson, Robert. *The Thinking Person's Guide to Climate Change*, second ed. Boston, MA: American Meteorological Society, 2019.

Kolbert, Elizabeth. *The Sixth Extinction: An Unnatural History*. New York: Picador, 2014.

LeCouteur, Penny, and Jay Burreson. *Napoleon's Buttons: How 17 Molecules Changed History*. New York: TarcherPerigee, 2003.

Losos, Jonathan B. *Improbable Destinies: Fate, Chance, and the Future of Evolution*. New York: Riverhead Books, 2017.

Miodownik, Mark. *Stuff Matters: Exploring the Marvelous Materials That Shape Our Man-Made World*. New York: Houghton Mifflin Harcourt, 2014.

Mukherjee, Siddhartha. *The Gene: An Intimate History*. New York: Scribner, 2016.

Orzel, Chad. *How to Teach Quantum Physics to Your Dog*. New York: Scribner, 2009.

Quammen, David. *Spillover: Animal Infections and the Next Human Pandemic*. New York: Norton, 2012.

—. *The Tangled Tree: A Radical New History of Life*. New York: Simon & Schuster, 2018.

Shubin, Neil. *Your Inner Fish: A Journey into the 3.5-Billion-Year History of the Human Body*. New York: Pantheon, 2008.

—. *Some Assembly Required: Decoding Four Billion Years of Life, from Ancient Fossils to DNA*. New York: Pantheon, 2020.

Stetka, Bret. *A History of the Human Brain: From the Sea Sponge to CRISPR, How Our Brain Evolved*. Portland, OR: Timber Press, 2021.

Yong, Ed. *I Contain Multitudes: The Microbes Within Us and a Grander View of Life*. New York: HarperColllins, 2016.

Zimmer, Carl. *A Planet of Viruses*, second ed. Chicago: University of Chicago Press, 2015.

—. *She Has Her Mother's Laugh: The Powers, Perversions, and Potential of Heredity*. New York: Dutton, 2018.

ACKNOWLEDGMENTS

Several people played important or even crucial roles in making this book a reality. The most pivotal of all was my literary agent, Luba Ostashevsky of Ayesha Pande Literary, because without her this book simply would not exist. Of the various agents I approached, she was the one who clearly saw potential in my proposal and sample chapters. She was also willing to take the time to help mold my proposal into something that was more interesting than what I originally had in mind. This meant a lot more work for me because I had to abandon my completed first draft (consisting of twenty-five short essays) and start essentially from scratch. But in the end it was definitely worth the extra effort, as her guidance resulted in a much better book. Equally important is that the reimagined proposal was much easier to sell to a publisher.

The second pivotal role in the creation of this book was played by Will McKay, the acquisitions editor at Timber Press who read the revised proposal Luba sent him and immediately took a strong interest in it. Then, for the next twelve months—as I researched, wrote, and submitted each chapter—Will worked closely with me, providing me with insightful feedback on how to improve the material I submitted. After I revised and resubmitted every chapter, Will arranged for a diverse group of scientists and science educators to review my work, providing me with additional insight as to how I might improve the details in each chapter.

Of the many reviewers Will lined up, I particularly want to thank Bret Stetka for his assistance on several of the life science chapters and Rhett Allain for his assistance on several of the physical science chapters. I greatly appreciated each instance in which Bret and Rhett proved to be sticklers for getting the details right, exactly the kind of feedback I had hoped to receive. Several other reviewers provided valuable comments on individual chapters, including John Doucet, Alie Caldwell, Michael Coe, Larry Scheckel, and Larry Olmsted, and I appreciated all of their insightful reviews.

After my year of working with Will, my completed manuscript was formally accepted and promoted to the next stage, to be guided by managing editor Michael Dempsey, who was also a joy to work with. At that point I was introduced to Lorraine Anderson, my copy editor, who performed the third pivotal role in the book's creation. Lorraine patiently polished every paragraph of the book, wrangling the wording and excising an amazing number of unnecessary words. I especially appreciated her many insightful suggestions on how I could improve the text—by adding bits of new material, reordering paragraphs, and trying new approaches when certain paragraphs just didn't work.

On the family and friends front, I benefited from the efforts of several people who read the entire manuscript and provided feedback from the standpoint of a typical reader. My wife, Marsha Rooks, read every chapter in its earliest stages, and my sister, Andrée Crow, read every chapter in its later stages. Both of them provided excellent suggestions for improvement. My longtime friend and colleague Steve Taffee also read the manuscript and provided me with several high-level recommendations to improve the book.

A few of the chapters in this book incorporate material originally published in my blog (The Philipendium), which can be found on the Medium website. However, the biggest part of each of these chapters consists of original material written specifically for this book, and most of the reused material has been significantly revised. I would like to express my appreciation to the readers of The Philipendium, whose comments and questions have encouraged me and given me helpful food for thought.

INDEX

A

adaptive radiation, 56, 58

adenine, 148

aerosols, 228, 230

agriculture, 240

AIDS (disease), 113–114, 249–251, 255, 259, 261

airborne molecules, and sense of smell, 17–18, 70, 217–218

air resistance, and gravitational acceleration, 40, 41

alcoholic beverages, 173, 175

algae, 21, 22

algorithms, 150, 151, 155, 157

alleles (variants), 51

alpha rays, 84–85, 87, 89

alternating current (AC), 190, 191

alveoli, 12

amino acids, 62, 147–149, 152–153, 166

ammonia, 175

amygdala, 210, 211, 218

analemma, 134–135

animals, 21–22, 68–69, 119, 165, 186

animal to human disease transmission, 249–251

antennae, 72

antibiotics
vs. antiviral drugs, 113

as magic bullets, 170, 171

overuse of, 105, 109, 118–119

and probiotics, 120

and viruses, 110

antibodies, 116, 145, 177

antigens, 177

antioxidants, 169–170

antiviral drugs, 113–114, 115, 171

apes, 53, 55–56, 60

aplastic anemia, 86, 87

apples, and gravity, 31–33

appropriateness, as survival factor, 46

aquatic animals, 75

archaeans, 61, 108

"the ascent of man" stereotype, 54–55, 58

Asian flu pandemic, 257–258

asteroids, 187

astronauts, 28–29, 30, 33, 39

astronomical unit (AU), 37

atmosphere, and IR energy, 233

atmospheric opacity chart, 101

atomic bombs, 193

autonomic processes, 215–216

avian flu (H5N1), 250

axial precession, 128

axial tilt, 130, 135

axis of Earth, 125–127, 130, 132, 133, 135, 138, 139

axons (parts of neurons), 212–213, 214–215

B

background radiation, 90–94

bacteria
and antibiotics, 118
and antibodies, 145, 177
evolution of, 61
killing of, 108–110, 113, 115–116
and spread of disease, 107, 260
and vitamin production, 176

bacteriophages, 113

Bacteroides organisms, 119

bad genes, 48

balance, sense of, 73–74, 78

basic reproduction number (R0), 248

bats, 69–70, 75, 250

Becquerel, Henri, 84

bees, 68, 75

benefits confused with causes, 14

beriberi, 164, 171

beta globin, 51

beta rays, 84–85, 87, 89

bile, produced by liver, 175, 176, 216

R. PHILIP BOUCHARD is a lifelong natural science nerd with a track record of creating successful educational media. As a software engineer and educator, he led the design of the famous 1985 computer game *The Oregon Trail*, along with other highly engaging products such as *Number Munchers*. Bouchard holds bachelor's and master's degrees in botany from the University of Georgia and the University of Texas at Austin. He frequently publishes fun, insightful educational essays on the natural sciences in *The Philipendium* on medium.com. When not engaged in his writing, Philip spends much of his spare time reading, exploring the world, photographing wildflowers, and collecting rocks.